T0191953

Sustainable Textiles: Production, Processing, Manufacturing & Chemistry

Series Editor

Subramanian Senthilkannan Muthu, Head of Sustainability, SgT and API, Kowloon, Hong Kong

This series aims to address all issues related to sustainability through the lifecycles of textiles from manufacturing to consumer behavior through sustainable disposal. Potential topics include but are not limited to: Environmental Footprints of Textile manufacturing; Environmental Life Cycle Assessment of Textile production; Environmental impact models of Textiles and Clothing Supply Chain; Clothing Supply Chain Sustainability; Carbon, energy and water footprints of textile products and in the clothing manufacturing chain; Functional life and reusability of textile products; Biodegradable textile products and the assessment of biodegradability; Waste management in textile industry; Pollution abatement in textile sector; Recycled textile materials and the evaluation of recycling; Consumer behavior in Sustainable Textiles; Eco-design in Clothing & Apparels; Sustainable polymers & fibers in Textiles; Sustainable waste water treatments in Textile manufacturing; Sustainable Textile Chemicals in Textile manufacturing. Innovative fibres, processes, methods and technologies for Sustainable textiles; Development of sustainable, eco-friendly textile products and processes; Environmental standards for textile industry; Modelling of environmental impacts of textile products; Green Chemistry, clean technology and their applications to textiles and clothing sector; Eco-production of Apparels, Energy and Water Efficient textiles. Sustainable Smart textiles & polymers, Sustainable Nano fibers and Textiles; Sustainable Innovations in Textile Chemistry & Manufacturing; Circular Economy, Advances in Sustainable Textiles Manufacturing; Sustainable Luxury & Craftsmanship; Zero Waste Textiles.

More information about this series at https://link.springer.com/bookseries/16490

Miguel Ángel Gardetti ·
Subramanian Senthilkannan Muthu
Editors

Handloom Sustainability and Culture

Entrepreneurship, Culture and Luxury

Editors
Miguel Ángel Gardetti
Center for Studies on Sustainable Luxury
Buenos Aires, Argentina

Subramanian Senthilkannan Muthu
SgT Group and API
Hong Kong, Kowloon, Hong Kong

ISSN 2662-7108 ISSN 2662-7116 (electronic)
Sustainable Textiles: Production, Processing, Manufacturing & Chemistry
ISBN 978-981-16-5969-0 ISBN 978-981-16-5967-6 (eBook)
https://doi.org/10.1007/978-981-16-5967-6

This Springer imprint is published by the registered company Springer Nature Singapore Pte Ltd.
The registered company address is: 152 Beach Road, #21-01/04 Gateway East, Singapore 189721,
Singapore

Preface

This title—Sustainability, Culture and Handloom—consists of three volumes—Entrepreneurship, Culture and Luxury; Artisanship and Value Addition; Product Development, Design and Environmental Aspects—which arises from an idea originated on August 7, 2020 on the occasion of the National Handloom Day in India. It is held annually to commemorate the Swadeshi Movement launched in 1905. It is a tribute that publishers do in order to celebrate this day.

Artisanship and craftmanship bring a whole new universe of small differences and dissimilarities in garments, that stand out in modern life and consumerist societies where garments appear to be all part of the same homogenized products. These pieces that hold the essence of heritage and uniqueness through techniques, materials and handmade might be defined as "cultural luxury". Beyond the essence and scarcity in handicraft products, these pieces are associated with an emotional value precisely due to the cultural process of maintaining cultural heritage based on traditions that are passed from generation to generation (Guldager 2015). Craftmanship luxury pieces carry the evidence of knowledge and ancient techniques that passes through generations, assuring cultural diversity and heritage, and those are intangibles of artisanship.

Publishers and authors of this title share both interests, and they are moved by the motivation to preserve and pass on the savoir-fair of those wise hands that ensure ancient techniques, passing this knowledge through generations and making luxury craftmanship possible.

Social and cultural aspects in craftmanship reflect and keep alive the essence and hold the values, the traditions and the cultural social exchange. The cultural sustainability of handicrafts is to keep this knowledge applied, and present, for example, to demonstrate through the daily use of handicrafts in our daily lives and be present of the essence of those pieces. In summary, envisioning and expanding the traditions and culture expressed in the artisan pieces assures, maintains and sustains cultural heritage and diversity (Na and Lamblin 2012).

This volume—*Entrepreneurship, Culture and Luxury*—consists of fourteen chapters that give a comprehensive outlook about this subject and begins with the work titled "Bapta Saree Revival: Reinventing the Past for a Sustainable Future" by

Swikruti Pradhan. In this chapter, the author investigates and deepens studies on Bapta Saree and Bapta weavers' communities to understand and unleash the potential of these pieces as commercial trading and cultural exchange to accelerate fashion sustainability.

The following chapter, "Community Entrepreneurship and Environmental Sustainability of the Handloom Sector" written by K. M. Faridul Hasan, Md. Nahid Pervez, Md. Eman Talukder, Sakil Mahmud, Vincenzo Naddeo and Yingjie Cai, explores the different aspects of environmental sustainability in the craft and fashion industries for associated handloom products. Highlighting on new markets, fair trade, decentralized distribution, employment creation, financial security, skill development, cultural integration (identity and diversity), innovative design and technology and community-based entrepreneurship as driving forces for sector expansion.

Then, Marisa Gabriel develop the chapter titled "HANDLOOMS: Unleashing Cultural Potentials" with research focused on handlooms in the Andes and the identity of weavers in the community. This chapter highlights the importance of handlooms as a symbol of the community's, specially of women's social interlace. It describes techniques and processes and explores the significance and potential of handlooms.

Subsequently, the purpose of Rana Alblowi's in the chapter "The Influence of Culture on the Sustainable Entrepreneur: An Investigation into Fashion Entrepreneurs in Saudi Arabia" is to explore the role that female entrepreneurs play in the field of luxury fashion industries and in striving to achieve Saudi Vision 2030 with the focus on sustainability and reviving fashion cultural heritage.

Moving on to the next chapter, "A Sustainable Alternative for the Woven Fabrics: "Traditional Buldan Handwoven Fabrics"", the authors Gizem Karakan Günaydın and Ozan Avinc present a general aspect and information about traditional Buldan weaves, Buldan handwoven fabrics, woven different textile products, handlooms techniques used in the town, the importance of weaving in the economic structure of Buldan district as well as unleashing the traditional hand weaving centers present in Buldan today.

The following chapter entitled "The Cultural Sustainability of the Textile Art Object" by Marlena Pop investigates cultural tools used by sustainable creative industries. It focusses on exploratory and experimental research in textile design, conducted with students of the Department of Textile Arts and Textile Design, to validate cultural tools and aesthetic material basis that allow to express through specific visual language, an entire individual universe, an archetypal heritage identity, with sustainable cultural values.

Then Ashna Patel develops the chapter titled "Conscious, Collaborative Clothing: A Case Study on Regenerating Relationships Within the Khadi Value Chain". The objective of this chapter is to analyze the birth of the khadi movement, its socio-economic and environmental impacts and the development in the khadi sector. Emphasis on the potential of collaborative relationships within the value chain to challenge the current speed and scale of design, production and consumer culture and to propose an alternative practice that aligns with planetary boundaries and human well-being opens up a path for designers to consider honing in developing hand-crafted luxury garments.

Later, Anna-Louise Meynell, in the chapter entitled "A Sustainable Model: Hand-loom and Community in Meghalaya, Northeast India", analyzes the integrated nature of handloom weaving in agricultural communities of Meghalaya (Northeast India) where the balance of the individual and the community has been maintained over generations. This paper explores the value of a part-time practice, aligned with the changing seasons and responsibilities of agriculture.

In the following, Meral Isler, Derya Tama Birkocak, and Maria Josè Abreu in the chapter "The Influence of Starch Desizing on Thermal Properties of Traditional Fabrics in Anatolia" explore three traditional fabrics in Anatolia namely Feretiko, Ayancik Linen and Burumcuk fabric focusing on thermal comfort properties.

In the next chapter "Indian Handloom Design Innovations and Interventions Through Sustenance Lens" V. Nithyaprakash, S. Niveathitha, and V. Shanmugapriya the authors overview the Indian handlooms in the current scenario, followed by reviewing the design attributes of the Kanjivaram, Ikat and Jamdani sarees, and a discussion on the changing face of Indian fashion semiology, analyzing innovation, direct and indirect implications of technology, the impact and the role of it in sustainable handloom products and Highlighting the effective design interventions in the manufacturing practice of the chosen Indian handloom products.

Then, in the chapter "How Translating Between Heritage and Contemporary Fashion Can Create a Sustainable Fashion Movement", the author, Dorothee Sarah Spehar, presents the different approaches brands and designers might have when working with artisan communities based on expert interviews. It analyzes the value of handcraft items with a focus on uniqueness, entrepreneurial business and social responsibility and analyzes the challenges of collaborating with artisan groups, implementing modernization and sustaining tradition to preserve cultural heritage.

Moving on to the next chapter, "Consumers' Attitudes Toward Sustainable Luxury Products: The Role of Perceived Uniqueness and Conspicuous Consumption Orientation", the authors, Andrea Sestino, Cesare Amatulli, and Matteo De Angelis, investigate the effectiveness of luxury brands' messages focused on product sustainability rather than on traditional luxury product features. The study sheds light on the role of perceived product uniqueness and consumers' conspicuous consumption orientation. Findings underline the role of conspicuous consumption in magnifying such effects and suggests different communication tools.

In the following, Karan Khurana in the chapter "Uzbekistan: The Silk Route of Handloom" presents the silk and cotton production in Uzbekistan and the research challenges and opportunities for the factories located in this region in the textile sector and provides viable solutions. It brings to light different issues that might serve for the stakeholders to initiate innovation and progress for a sustainable textile future.

Finally, Claire Shih, in the chapter entitled "Aesthetic Capitalism and Sustainable Competitiveness in Urban Artisanal Networks", intents to explore through case studies the ethics of sustainable development and the UN's Agenda 2030, in relation to those of aesthetics in its capitalist context and also explore empirically the emergence of a counter current to the luxury world, detached from the ubiquitous luxury brands and their global domination. Presenting this small subsector of craft-oriented luxury and its potential to lend support to the UN's concepts of sustainable development and improvement of the city's quality of life.

Buenos Aires, Argentina   Miguel Ángel Gardetti
Hong Kong, Hong Kong   Subramanian Senthilkannan Muthu

References

Guldager S (2015) Irreplaceable luxury garments. In: Gardetti MA, Muthu SS (eds) Handbook of sustainable luxury textiles and fashion, vol 2, Springer, Singapore, pp 73–97

Na Y and Lamblin M (2012) Sustainable luxury: Sustainable crafts in a redefined concept of luxury from contextual approach to case. Stud Making Futures J. 3, ISSN 2042–1664

Contents

About the Editors

Miguel Ángel Gardetti (Ph.D.), founded the **Centre for Study of Sustainable Luxury**, the first initiative of its kind in the world with an academic/research profile. He is also the founder and director of the "Award for Sustainable Luxury in Latin America". For his contributions in this field, he was granted the "**Sustainable Leadership Award** (academic category)," in February, 2015 in Mumbai (India). He is an active member of the **Global Compact** in Argentina—which is a **United Nations** initiative—and was a member of its governance body—the Board of The Global Compact, Argentine Chapter—for two terms. He was also part of the task force that developed the "**Management Responsible Education Principles**" of the United Nations Global Compact. This task force was made up of over 55 renowned academics worldwide pertaining to top Business Schools.

Dr. Subramanian Senthilkannan Muthu currently works for SgT Group as Head of Sustainability and is based out of Hong Kong. He earned his Ph.D. from the Hong Kong Polytechnic University and is a renowned expert in the areas of Environmental Sustainability in Textiles and Clothing Supply Chain, Product Life Cycle Assessment (LCA) and Product Carbon Footprint Assessment (PCF) in various industrial sectors. He has five years of industrial experience in textile manufacturing, research and development and textile testing and seven years of experience in Life Cycle Assessment (LCA) and carbon and ecological footprints assessment of various consumer products. He has published more than 100 research publications, written numerous book chapters and authored/edited over 100 scientific books in the areas of Carbon Footprint, Recycling, Environmental Assessment and Environmental Sustainability.

Bapta Saree Revival: Reinventing the Past for a Sustainable Future

Swikruti Pradhan

Abstract The handloom textile craft of Odisha is preserving the essence of its legacy and culture as on today, the incredibly beautiful state of in Eastern India. The sacred land blessed with an extensive natural beauty and abundant cultural heritage is globally recognized for its rich traditional handloom and handicrafts. Odisha is known for the origin of various handloom-weaves with the names as the places of origin. The weaves are highly influenced by history and Hindu mythology, especially the Lord Jagannath culture. They often showcase temple borders, mythological designs, and the traditional colors of the Lord. One of such famous sarees is Bapta, an indigenous amalgamation of silk and cotton with remarkable comfort and lustre. Unfortunately, the Bapta sarees slowly vanished over time, as the weavers could not seize the right market for them. Interestingly, decades later, recently, there have been traces of Bapta sarees being woven in a few parts of Chhattisgarh (the neighboring state of Odisha), showing a testimony of the shared history of cultural exchange. It is believed that the collaborative weaving skills of both states had a common geographical identity in the past without any comprehensive written documentation. Local start-up design labels like Rustic Hue are putting their efforts to revive this age-old craft most genuinely through their 'Bapta Revival Project' by working with the Kosta and Bhulia Meher communities in Western Odisha. Sustainable development of a community is not possible barring culture as it shapes identity. This chapter documents as well as reviews the revival of Bapta during its journey.

Keywords Bapta saree · Craft revival · Communities · Local practices · Cultural sustainability

S. Pradhan (✉)
Rustic Hue, Bhubaneswar, Odisha, India

© The Author(s), under exclusive license to Springer Nature Singapore Pte Ltd. 2021
M. Á Gardetti and S. S. Muthu (eds.), *Handloom Sustainability and Culture*,
Sustainable Textiles: Production, Processing, Manufacturing & Chemistry,
https://doi.org/10.1007/978-981-16-5967-6_1

1 Introduction

The textile craft of Odisha is preserving the essence of its legacy and culture as on today, the incredibly beautiful state of East Coast, India. Handloom weaves have always been significant to the abounding tradition and heritage of the state as well as the cultural symbols of the nation. These handloom weaves are inimitable calls for a high skillfulness. Blessed with an extensive picturesque landscape and rich cultural heritage, Odisha is globally recognized for its rich traditional handloom and handicrafts. The state has the immortal history of changing the great emperor and the cruel ruler to a philanthropic king named Ashoka from 'Chandasoka' to 'Dharmasoka' and even the name of the artisans who are the scapegoats, reflected during modern ruled British-India. However, the legacy of handlooms and handicrafts serve as a backbone of economic empowerment and employ to the less educated and poor in the rural sector as on today. The rhythms, steady sounds of shedding, picking, and beating denote the prominence of the handloom industry and its contribution to the economy of the state. The weavers have transformed handloom weaving and weave from just passion to actual means of economy over the ages unconsciously by obtaining basic business intelligence [1].

Until now, the yarn tie-dye art and manually handwoven sarees remain unmatched, unparalleled and far away from the reach of industrial automation. The weaves are highly influenced by history and Hindu mythology, especially the Lord Jagannath culture. They often showcase temple borders, mythological designs, and the traditional colors of the Lord. One of such famous sarees is Bapta, an indigenous amalgamation of silk and cotton with remarkable comfort and lustre. They have neither matte like cotton nor too shiny like silk texture, produced in a very few districts of Odisha by different weavers' communities; less expensive than the traditional tussar and mulberry silk sarees yet elegant. Unfortunately, the Bapta sarees slowly vanished over time, as the weavers could not seize the right market for them. Interestingly, decades later, recently, there have been traces of Bapta sarees being woven in a few parts of Chhattisgarh (the neighboring state of Odisha), showing a testimony of the shared history of cultural exchange. It is believed that the collaborative weaving skills of both states had a common geographical identity in the past without any comprehensive written documentation. Local start-up design labels like Rustic Hue are putting their efforts to revive this age-old craft most genuinely through their 'Bapta Revival Project' by working with the Kosta and Bhulia Meher communities in Western Odisha. Sustainable development of a community is not possible barring culture as it shapes identity.

This chapter extensively investigated for the insights with qualitative research methods including desk search such as textual and visual content analysis and field visits, both participant and disguised observations, interviews, and interactions with the Bapta weavers' communities, the lifestyle of the weavers and how the revival of Bapta sarees can be a potential drive for both commercial trading and cultural exchange between two neighboring states achieving fashion sustainability.

Fig. 1 The Meher community

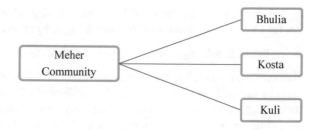

2 Bapta Saree

i.   Origin and History

The handloom industry of Odisha has a long history with various weaves in local regions such as Sambalpuri Ikat, Bomkai, Habaspuri, Kotpad, Khandua, Siminoi, and Bapta. The designs, traditional techniques, quality, and genuineness have been complimented worldwide. Bapta sarees are believed not to be more than half a century old. The history dates back to the time when handwoven mulberry silk sarees with ikat or baandha [2][1] (yarn tie-dyed) and tussar sarees woven by Meher Bhulia and Kosta communities in Western Odisha were well known like today but could be afforded by a niche group of customers. According to Shri Prahlad Meher, the Secretary of Dewangan Weavers' Co-operative Society in Remunda, the society was formed in the year of 1984, which has 81 registered weavers now out of them 25 are regularly active. According to him, Late Shri Srinibas Meher first conceptualized the Patta Bapta saree in the same year. His idea was to come up with a type of saree in a price range in between mulberry silk (Patta) and cotton (Suta) sarees. Mulberry silk, Tussar, and Cotton are the major natural yarns used in the handloom in Odisha. The uniqueness of the saree is the amalgamation of mulberry silk (Patta)/tussar warp and cotton weft or vice versa. Thus, the saree looks subtle in its texture and appearance instead of lustrous like mulberry silk (patta) saree or matte like cotton.

ii.   The Weavers Communities

Handloom weaving is a caste based profession and usually based on traditional skill sets that require manual labor and produces weaves in small batch size [3]. In Odisha, the Meher community is dominantly seen in the Western part and has been divided into three classes—Bhulia Mehers, who have expertise in the art of yarn tie and dye (*Ikat* weaving in cotton and silk; Kosta Mehers, who work mostly with tussar yarns and three-shuttle loom and do not follow the traditional *Ikat* technique; and Kuli, who are labor class Mehers with little skills [4].

The beautiful Bapta sarees are generally woven by two different groups of the Meher community (Fig. 1). The yarn tie-dyed (*Ikat*), which is essential for the aanchal part or sometimes incorporated in the body of the saree is prepared by the Bhulia

---

[1] https://grandmaslegacy.wordpress.com/tag/bapta/.

Meher weavers communities and then sent to the Kosta Meher weavers who complete the final weaving on three shuttle loom using temple borders (*Phoda Kumbha*).

iii.   Traditional Designs and Cultural Significance

Woven on a pit loom, Bapta is a uniquely wonderful saree that results from the confluence of two different yarns—Tussar or Mulberry Silk and Cotton. In the simplest form, Bapta can be explained as an indigenous amalgamation of silk and cotton with remarkable comfort and lustre. It is a beautiful amalgamation of silk warp and cotton weft (or vice versa) interwoven into each other with temple borders (*Phoda Kumbha*) and yarn tied-dyed (*Ikat*) aanchal and simple *Ikat* motifs on the body (Fig. 2). The borders are often seen in contrast colors with the aanchals having intricate ikat work (Fig. 3).

Traditionally, the two borders across the body of the saree were being woven with temple borders (*Phoda Kumbha*) using 3-shuttle weaving technique. The body of the saree used to has small motifs called *Buta* woven as extra weft on the loom (Fig. 4). Some sarees used to have both *Ikat* and Temple borders (*Phoda Kumbha*) and *Ikat* aanchal of comparatively smaller width (Fig. 5).

The weavers have been surrounded by calm and lush green nature, which reflected their work on the Bapta sarees they used to weave. The motifs of the sarees were mostly inspired from nature and surroundings in circular and oval shapes, giving a sense of sheer elegance. The weavers used to compose designs with wonderful

Fig. 2 Two decades old *Rajni* Patta Bapta saree (*Source* Prahlad Meher's wife)

Fig. 3 Intricate *Ikat* aanchal

Fig. 4 A 15 years old Patta Bapta saree with *Buta* on the body (*Source* Kundan Pradhan)

concepts taking inspirations from immediate environment such as plants, vegetables, and animal creatures. There are examples of some classic traditional motifs such as kalara (bitter gourd), the atasi flower, the kathi phool (small flower), maachi (fly), rui

Fig. 5 A 35 years old Patta Bapta saree (*Source* Kundan Pradhan)

maach (carp-fish), kainchha (tortoise), padma (lotus), mayura (peacock), and chadhei (bird) woven into the aanchal and body of Bapta sarees.

The motifs and designs depict elements from nature. Unlike most of the common forms of motifs that are woven on the face of various textiles, Bapta is definitely different. Inspired by folklore that grows from the roots of the Odia culture, Bapta was an emerging saree not only in the Western part but the entire state. The weavers use the motifs as a form of artistic expression. But, there should be a distinction between craft and a form of art to protect the price points, categories, and careers of the artists and craftsmen [5].

The highlight of the Bapta sarees is the borders with detailed *Phoda Kumbhas* (temple motifs), which are woven using 2 or 3 shuttle looms and two weavers are engaged for the finishing of the saree weaving. There are also instances of the sarees featuring motifs that are inspired by tribal art such as Rudraksha in the borders.

Colors were inspired by the effect of contrast, generally the body of the saree in different shades and tones of Indian Sandalwood. An equally absorbing yet mesmerizing dark magenta, blue, maroon, purple, or black were seen. Sometimes, the *ikat* weaves in aanchal were arranged in vertical lines or the aanchal might be double shaded, giving out a reflection of a regal look. Latticework creating small diamond-like shapes also used to be commonly found on the border of a Bapta saree.

However, Bapta sarees could never gain great attentions globally or in form of different products viz yardages, stoles, scarves, and attires like other renowned handloom weaves such as the Bomkai, the Sonepuri, or the Sambalpuri Ikat.

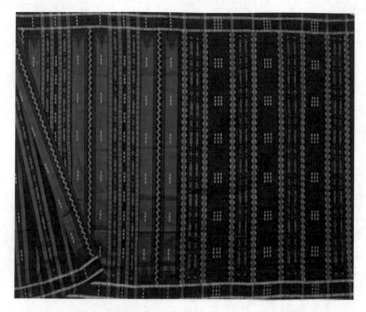

Fig. 6 Patta Bapta saree with Ikat and extra weft motifs in vertical manner (*Source* Manoj Meher, Remunda)

iv. Re-imagining Traditional Designs

The Bapta Sarees have evolved all these years in terms of color combinations and designs. Although the production of these sarees by the Kosta weavers in the Bargarh district have reduced drastically, the sarees are being woven in the same template of design for more than a decade now without much changes in the pattern. The weavers are using cotton warp and have reduced the part of mulberry silk in the weft replacing it with cotton to decrease the overall cost of production of the saree (Figs. 6 and 7).

Handloom weaving has been a community activity and constant source of income for generations now. Handloom clusters have witnessed many design intervention programs and collaborative projects with designers and the government agencies. It is often observed that many handloom weaves have not received as much recognition in their home state as in foreign lands. There is a lack of market exposure for which the weavers themselves are limited to traditional designs. Sometimes they are reluctant to join skill-development training programs organized by the government to enhance their skills and explore various market opportunities and therefore they depend on external support reducing themselves to handloom laborers. Proper design intervention either through the help of the Government or private labels can enable the weavers to progress toward a sustainable livelihood and explore markets for themselves [6].

However, the effort of design intervention by local design label Rustic Hue is bringing in the much needed empathetic understanding and holistic vision to connect

Fig. 7 Patta Bapta saree with wide Phoda Kumbhas, Ikat, and extra weft motifs in vertical manner (*Source* Monoj Meher, Remunda)

and integrate the these endeavor toward a positive result to some extent. The designs and color combinations are fresh and have a contemporary take on this dying craft (Figs. 8 and 9). The label experiments with the size and shapes of the *Phoda Kumbhas*, the visual texture of the body, and the detailing in the aanchal.

v.   Gradual Extinction of the Craft

The beautiful handloom saree has been struggling hard to survive in today's time. The local communities across different villages in the district have decreased weaving this particular saree leading to its extinction. It is believed that they failed to trap the right market and audience for the saree. Lack of opportunities and appeal to a contemporary audience took this saree in the verge of extinction. After hectic travels, interviews, and thorough research, many shades to the story were revealed. The Kosta Weavers' Co-operative Society Ltd is one of the oldest societies with all Kosta weavers in the village of Barpali in Bargarh district established in 1954 (Fig. 10).

The Kosta weavers in the village used to produce Tussar Bapta Sarees with cotton warp and tussar weft with *phoda kumbha* using three shuttle loom and aanchal with *ikat* yarns sourced from Bhulia weavers. According to Sankhamani Meher, an active member of the society, the Bapta sarees witness a major barrier due to some of the weavers groups in the villages who decreased the part of the tussar resulting in lesser costs of the sarees and sold off to the middle-men or merchants at lesser price points, for which, the authentic Bapta sarees once conceptualized by the great artisans of the previous generation lost its real charm. Some weavers started procuring impure

Fig. 8 Patta Bapta saree in acquamarine-butter color with stripes detailing (*Source* Rustic Hue)

Fig. 9 Heena green Patta Bapta saree with Ganga-Jamuna borders (*Source* Rustic Hue)

poly mix tussar or mulberry silk (Fig. 11) to reduce the base cost of production to even greater extent that the less aware consumers shifted toward buying the cheaper Bapta sarees instead of the authentic ones.

Santkabir Awardee Shri Khsetra Mohan Meher gave an insight to the extinction of the saree that the majority of the Kosta weavers in Barpali village did not want to

Fig. 10 Kosta Weavers'
Co-operative Society Ltd in
Barpali, Bargarh (*Source*
Rustic Hue)

Fig. 11 Manmade impure
tussar yarns produced in the
mill (*Source* Rustic Hue)

be unethical with this practice, yet it was commercially uneconomical for them to continue to produce the authentic Bapta sarees due to the pressure of competition in the market. The Kosta weavers in this Co-operative society have been approached by many yarn producers and suppliers time to time to use the impure tussar or mulberry silk, but they have strictly been away from using any manmade fiber. When a major chuck of the community stop producing because of their ethical values and other reasons, it is evident that why the saree was on the verge of extinction. To top it up, their own community sometimes gets in cutthroat competition and mixes impure yarns and sells it at much cheaper rates. Apart from all these, the main barrier is the middlemen and organizations; forums that claim to "promote Indian craft". These organizations or middlemen purchase handwoven sarees from the weavers and sell at unimaginably higher prices, thereby making the saree unaffordable.

They face competition not just from the retailers/merchants who are selling them off to the less aware customers but also from our fellow community who choose to be unethical. It is unfair on hard work of other weavers in the community.

The Meher Bhulia and Kosta weavers do not have a documented form of Bapta sarees. There is not much evidence of textual or visual documentation of this saree as such. Like many traditional crafts Bapta saree has also become memory based, and only memory is the knowledge bank of the weavers' traditional methods and materials [7]. The weavers need encouragement, exposure, and support to be educated about such practices.

The handloom industry is labor intensive with a legacy of unparalleled craftsmanship. Handloom products are not only a part of daily wear of the locals but also a weakness for the elite in India and abroad. While many museums and galleries across the cities in the country showcase the work of urban artists or researchers practicing traditional art forms, none of them have documented or showcased this saree form. But, the most of their work have the upper class urban audience. While it can help preserve and advocate the craft form and also encourage the struggling rural artisan to keep continuing the legacy.

While the growing demand for power loom products are affecting the cottage industries and creating hindrance in their growth, the beauty and brilliance of handloom products of Odisha have potential to create large employment opportunities and attract the foreign markets [8].

3 Cultural Exchange and Social Sustainability

Cultural exchange means the sharing of different ideas, traditions, and knowledge with someone who might have a completely different background than the other one. Cultural exchange has scopes for the world to become a more beautiful, connected place for everyone. It becomes easier to understand someone from a different community with social background better if one recognizes the intrinsic value that each member of that community holds. Better understanding about one another can help work together for impactful change.

The Bapta sarees in the recent past have seen traces in eastern side of Chhattisgarh (the neighboring state of Odisha), where a group of Kosta weavers' communities reside with the surname Dewangan and practice the same weaving techniques as the Kosta Mehers in Western Odisha. They have started procuring the *ikat* yarns from the Bhulia Mehers from Western Odisha and combining them with Kosa yarns (Desi

I have draped Bapta woven in Sambalpur way back in early 80s. With passage of time my Odisha Baptas frayed like Chanderis. Slowly they vanished from the market. After a long time I saw Bapta in #kosaman's shop in Bhilai. He said they were woven in Chhattisgarh and not in Odisha. It was just the opposite of what I had in the past. Whether woven in Odisha or in Chhattisgarh it is a beautiful marriage of two states #2statesin1saree. Chhattisgarh weavers procure the finished cotton Bandha weft (Ikat) threads from Bargarh in western Odisha. The warp is kosa from Chhattisgarh. It's woven in Champa by Chhattisgarhi Devangan weavers incorporating the ikat threads into their Kosa.

Liked by sari_teller and 665 others

DECEMBER 7, 2019

Add a comment...   Post

Fig. 12 A post by Vijayalaxmi Chhabra on Instagram in her Bapta saree form Chattisgarh

Tussar) to produce Bapta sarees (Fig. 12).[2] This opens the gateway to trading and cultural exchange between two neighboring states across the borders.

India may seem diverse and different, if you travel by train or airplane but if you walk across the boundaries of states you can see vast similarities between the cultures, traditions, folk dances, songs, paintings, art forms, and languages. The border areas of two states represent a subtle blend of the seemingly contrasting cultures, so much so that, when you walk through the borders you might not realize the difference at all [9].[3]

Cultural exchange has the power to thrive in a new environment and shows the significance of similarities rather than differences, highlights the beauty of diversity, and makes the world a more beautiful, connected place for everyone [10].[4]

Without the support of collaborations, success of any innovation is not possible. NGOs and non-profits have assisted in bringing together resources, mobilizing people, and volunteers, overcoming language barriers, etc. Partnering with more nonprofits has more impact on their beneficiaries adding to their project's growth.

Commercialization of textiles is a great concept, given that it respects and keeps the core ideology and appeal of the craft culture intact. The instances of cultural exchange

---

[2] https://www.instagram.com/p/B5wc_VBJfor/?hl=en.

[3] https://www.youthkiawaaz.com/2016/06/handlooms-sustainable-clothing/.

[4] https://greenheart.org/blog/exchange/cultural-exchange-why-it-matters/.

through traditional textiles and specifically the Bapta sarees between Odisha and Chattishgarh might open a wide range of opportunities for the Meher communities spread across both the states. There can be collaborations between designers and weavers to revive the dying weave, and the consumers will also be aware of more materials now beyond just their visual and sensual appeal, but also with their origin and importance of existence. The present generation of consumers is also curious about how things are made and who made them.

It would be great for the weavers in the rural villages to get exposure to the right kind of market and audience rather than leaving behind by market forces, rapid development of Power loom on an extensive scale [11].

Self-sustainability is becoming more important with the energy and environmental crisis. Handlooms are environmentally friendly and are an independent and autonomous in technology. The industry presents a quality on itself to sustainable development and reduction of negative impacts on environment and ecology as well as providing sustainable employment for rural India. The handloom is an ancient industry in India, and its use varies throughout the region; it has become a mature industry, while in some parts of India it is still used primarily as a household staple. Handloom production is a family-based activity and is inseparable from the living atmosphere of the handloom families. Handloom sector constitutes a distinct feature of the rich cultural heritage of India and plays a vital role in the economy of the country.

4 Craft Revival for a Sustainable Future: An Example of 'Bapta Revival Project' by Label Rustic Hue

The responsibility of a designer goes beyond glamour and trends. More often than not, craftsmen cannot release information about their own work done in association with a brand. But designers massively influence trends. So if something is popular among the trendsetters, it is going to influence the buying capacity of the masses, and that is where a designer can majorly contribute to the crafts community.

The Bapta Revival project by Rustic Hue focuses on design intervention to revive this weaves form, which is almost on the verge of extinction[12].[5] The sarees have a contemporary take on the existing craft with minimal or unusual color combinations, design patterns, and Phoda Kumbha styles (Fig. 13). This reflects the enormous work done by designers and the Meher community with a collaborative approach to ensure that crafts remain key to our experience of culture.

While craft conservation includes documentation of a traditional craft, its significance, and history, the revival of it is through creative and product development workshops with the weavers on dyes and processing. The label encourages the dialogue between the two representations of society with different communities involved in one weave form. The label put effort to make the weave form to find position in

---

[5] http://cms.newindianexpress.com/magazine/2019/feb/03/old-is-new-and-natty-1933244.html.

Fig. 13 A subtle Fern Green
Patta Bapta saree with Pale
Green borders with Phoda
Kumbhas and Sambalpuri
Ikat Aanchal (*Source* Rustic
Hue)

the dynamic handloom culture globally. It is trying to create space for this dying handloom weave rich with stories about the beautiful communities and traditional techniques. The label has been voicing about their revival project to bring awareness among the consumers on the social media (Fig. 14).[6]

The hard work in reviving the saree forms have been appreciated and getting good responses in the market so far. It has also caught attention of celebrities like Archita Sahu, an Odia film actress, who has worn Patta Bapta saree by Rustic Hue in different programs or events and also voicing about the craft form on her social media (Figs. 15[7] and 16[8]).

The label has been able to spread the message of Bata saree, the weave form, and culture among the urban audiences. At the rural level, it works with weavers of Kosta and Bhulia Meher communities [14].[9] The work happens in collaboration with the weavers and their communities to understand the intricacies of their lives, their cultural heritage, and the challenges they face in practicing their craft. Interacting with the communities helps the designer's team to understand various social issues they face—like education for their children and health services. The label believes there should be an approach to encourage the weavers to incorporate their culture

---

[6] https://www.instagram.com/p/CASGUGMJcIc/.

[7] https://www.instagram.com/p/CGFhOQiJ-6p/.

[8] https://www.instagram.com/p/CHStaNopZb0/.

[9] http://www.odishabytes.com/unfolding-the-rustic-charm-of-odishas-baptasaree/.

studio.rustic.hue

The picture is of the beautiful 'Patta-Bapta' saree, a unique combination of cotton and mulberry silk yarns with traditional temple borders (Phoda Kumbha - woven by two weavers) and #sambalpurikat aanchal. They are the easiest and super comfortable to wear for longer durations. It's hard to find authentic 'Patta-Bapta' sarees these days. The saree was in the verge of extinction as hardly two weavers families were left doing this regularly. We spent a lot of time not only on research to revive this age old saree with contemporary designs in terms of motifs and textures, but also to persuade the talented weavers to work on it regularly to bring it back in demand. The whole experience has been challenging yet exciting.

Liked by saree_lovestories and 28 others

MAY 17, 2020

Add a comment...   Post

Fig. 14 A post on Bapta Revival Project by Rustic Hue on Instagram

archita sahuofficial ● • Follow

architasahuofficial ● My love for #sustainablefashion ☺Here is a unique combination of cotton and mulberry silk yarns with traditional temple borders (Phoda Kumbha - woven by two weavers) and #sambalpurikat aanchal. It's hard to find authentic 'Patta-Bapta' sarees these days. The saree was on the verge of extinction as hardly two weavers families were left doing this regularly. A lot of time was spent only on research to revive this age old saree with in terms of motifs and textures, to bring it back in demand. M glad I could get this customised.
It's an urge to everyone to be #vocalforlocal.

Bapta Revival Project by @studio.rustic.hue

Liked by sneha1287 and 26,194 others

OCTOBER 8, 2020

Add a comment...   Post

Fig. 15 A post by Archita Sahu on Instagram in Rustic Hue's Patta Bapta saree for the reality show Dance Odisha Dance on Zee Sarthak TV [13]

architasahuofficial ● • Follow ···

architasahuofficial ● Wearing a Patta Bapta Saree - a revived handwoven saree from the "Bapta Revival Project" that was on the verge of extinction as the local weavers communities gradually decreased weaving this. Paired up with a beautiful mulberry silk crop jacket from the "Jugaad" collection, up cycled out of post production left over fabrics in the studio preventing them to go on land filling. It's high time for us to care for the environment and also be #vocalforlocal. 🌿
@studio.rustic.hue
@swikruti.pradhan
Gold jewelry from the AMEYAA collection of @arundhatijewellersofficial

Liked by **soundarya_ms** and **33,936 others**
NOVEMBER 7, 2020

Add a comment... Post

Fig. 16 A post by Archita Sahu on Instagram in Rustic Hue's Patta Bapta saree for the Collection launch event of Arundhati Jewllers

into the saree they make for the urban market, which makes the products personal for them, and introduces the customer to the cultural narrative of their community.

5 Bapta Saree—A Form of Sustainable Clothing

As a handloom product, bapta sarees are produced using pit looms and no harmful chemical dyes making it eco-friendly. This ethnic product not only has a cultural dimension but also has a great employment opportunity in the community. The cultural dimension of sustainability is the link among other dimensions of sustainable development such as social, economic, and environment. Preservation of cultural identity of a community leads to cultural sustainability [4] and hence ensures the saree to become culturally and environmentally sustainable.

The increasing population on the earth has huge impact on human environment. As conveyed by the United Nations Conference on Environment and Development, healthy practices are the need of the hour for achieving sustainable development. The socio-economic pillars for sustainable development must not be ignored [15].

Bapta sarees without realization of the community itself are indisputable part of the culture and economy. If the weavers' communities do not realize their potential and future with this saree, then it is imperative to work toward their livelihood and the craft's sustenance [16].[10]

Handloom and handcrafted products should become a part of everyone's everyday wardrobe. Each product is a reminder that what a pair of hands, dipped in love, using ancient techniques can create can never be replicated on a machine. Making each one of its kinds a treasure forever. Even though one cannot progress without technology at the moment, but it is necessary to understand that excessive dependency on power loom products and complete negligence toward hand-made and sustainable products could lead us toward a doomsday of its own.

It is essential to value our surroundings and traditional handloom weaves for future growth and prosperity, as unsustainable practices cannot be beneficial in the long run although they benefit immediately [17].

6 Conclusion

Handlooms symbolize the legacy of a rich culture, and they should be valued. The contribution of the handloom sector to the rural economy and society is crucial. The traditional textiles have always brought glory to the rich diverse culture and heritage of the country. Hence, it is important to preserve and protect the skills and knowledge of traditional textiles. A wide awareness of the endangered arts and crafts will help them revive and bring a new life altogether [18].[11]

Craft revival focuses on recognizing a weaver community's culture, skills, and extending the same to the masses. It strengthens the weaver's abilities to innovate, explore market places, and exclude the middlemen. Collaborative projects among designers and weavers bring a lot of scopes of innovation, expansion of the weave vocabulary, and reestablishment of dying craft traditions and tapping contemporary markets [19

It is important to understand the need of preservation of a dying weave form; Bapta saree represents the skill exchange between two groups of a whole community it takes years of practice and dedication to learn a craft. It is unfortunate that lack of creative and market opportunities, exploitation by the middlemen, industries, and designers leads to most artisans switching occupations to odd jobs that pay very less. Craft and craftsmanship need attention to bring about effective sustainable development through policies [20].

Speaking of solutions, the upcoming artisans need to be educated more about the crux of local practices, culture, and the value at the global stage and its impact on their future generations. The artisans of these dying crafts and textiles need evolution to remain alive. Just like anything else in the universe, in order to survive, an artisan's skills need to involve, adapt, and evolve, and for them to be able to do that, we must stand together to provide the support they need [21].

---

[10] https://www.thehindu.com/entertainment/art/crafting-a-new-future/article26331025.ece.

[11] https://www.thehindu.com/entertainment/art/crafting-a-new-future/article26331025.ece.

There is an urgent need of consumer awareness. A consumer holds more power than one can imagine. The buying capacity influences everything around us, from trends to a company's CSR (Corporate Social Responsibility). To begin with, opting for sustainable options can be the beginning of something vastly impactful.

Handloom weaving is specialized activity as it is intricate and complex and requires the entire communities to specialize in the production of a certain craft product. It will be challenging in the future to find ways and means to preserve and nurture design specialization, skill, and artistry to ensure Indian craft regains its unique position in the world.

'Bapta' sarees, which are lesser known among the present generation but were once massively popular among the previous generation of women consumers, have still hope to gain back the splendor. Six yards of clothing witnesses the touch of multiple human hands through many stages of various processes of two different groups of a community. It is needless to say that this saree not only has the potential to portray rich culture and a form of sustainable clothing, showcase the beauty of complex and labor intensive weave, but also keep the sub groups of a weaver's community unit and provide economic benefits.

The future of the revival of this saree is definitely going to be a glorious one in many aspects. The native people also have the responsibility of bridging the gap between the weavers and patrons so that no middleman pointlessly pilfers the profit and leaves them with something that does not even cover his basic price. It is essential to support and empower them so that they can continue their craft with pride. It is amazing to see how the culture and tradition has been passed on to the next generation over centuries, which also leaves the generation with responsibilities of becoming the custodians of the rich cultural past. The generations to come, both the craftspeople and the consumers, should not miss out on this saree weave, which has much more to give beyond a rich cultural background.

References

1. Vijayshree PT, Hema B (2011) The raising era of artrepreneurship—a case study on saipa. Annamalai Int J Bus Stud Res 3(1):p 24–29
2. Kekkar K (2017) https://grandmaslegacy.wordpress.com/tag/bapta/. Last accessed 15 Jan 2021
3. Chick A, Micklethwaite P (2011). Design for sustainable change: how design and designers can drive the sustainability agenda, vol 38. AVA
4. Pradhan S, Khandual A (2020) Community, local practices and cultural sustainability: a case study of sambalpuri ikat handloom. In: Muthu G (ed) Sustainability in the textile and apparel industries. Springer, Cham, pp 121–139
5. Risatti H (2007) A theory of craft: function and aesthetic expression. The University of North Carolina Press, America
6. Botnick K, Raja I (2011) Subtle technology: the design innovation of Indian artisanship. Des Issues 27(4):43–55
7. Kapur H, Mittar S (2014) Design intervention & craft revival. Int J Sci Res Publ 4(10):1–5
8. Vyshak A, Athira V, Anandavalli T (2018) Development with sustainability: a study of small scale sector with special reference to handicrafts and handlooms). Int J Pure Appl Math 119(16). https://www.instagram.com/p/B5wc_VBJfor/?hl=en. Last accessed 7 Oct 2020

9. Rai N (2016) Handlooms sustainable clothing. https://www.youthkiawaaz.com/2016/06/han dlooms-sustainable-clothing/. Last accessed 16 Jan 2021. https://greenheart.org/blog/exc hange/cultural-exchange-why-it-matters/. Last accessed 16 Jan 2021. https://www.instagram. com/p/CHStaNopZb0/. Last accessed 15 Jan 2021

10. https://greenheart.org/blog/exchange/cultural-exchange-why-it-matters/

11. Balaji NC, Mani M (2014) Sustainability in traditional handlooms. Environ Eng Manag J (EEMJ) 13(2)

12. Sahu D (2019) Old is new and natty. http://cms.newindianexpress.com/magazine/2019/feb/03/ old-is-new-and-natty-1933244.html. Last accessed 7 Dec 2020

13. https://www.instagram.com/p/CGFhOQiJ-6p/. Last accessed 15 Jan 2021. https://www.instag ram.com/p/CASGUGMJcIc/. Last accessed 15 Jan 2021

14. Mishra S (2018) Unfolding the rustic charm of Odisha's Bapta Saree. Odisha Bytes, 7 August 2018. http://www.odishabytes.com/unfolding-the-rustic-charm-of-odishas-baptas aree/. Last accessed 7 Dec 2020

15. Mahapatra SK, Ratha KC (2017) Paris climate accord: Miles to go. J Int Dev 29(1):147–154

16. Venkatraman L (2019) Crafting a new future. The Hindu. https://www.thehindu.com/entertain ment/art/crafting-a-new-future/article26331025.ece. Last accessed 12 Dec 2020

17. Khandual A, Pradhan S (2019) Fashion brands and consumers approach towards sustainable fashion. In: Muthu (ed) Fast fashion, fashion brands and sustainable consumption. Springer, Singapore, p 37–54

18. Mahapatra A (2019) Revival of dying Indian art and crafts. https://www.atg.world/view-art icle/28753/revival-of-dying-indian-art-and-crafts. Last accessed 15 Jan 2021

19. Temeltaş H (2017) Collaboration and exchange between "Craftsman" and "Designer": symbiosis towards product innovation. Des J 20(sup1):S3713–S3723

20. Mignosa A, Kotipalli P (Eds) (2019) A cultural economic analysis of craft. Palgrave Macmillan

21. Soini-Salomaa K, Seitamaa-Hakkarainen, P (2012) The images of the future of craft and design students–professional narratives of working practices in 2020. Art, Des Commun High Educ 11(1):17–32

Community Entrepreneurship and Environmental Sustainability of the Handloom Sector

K. M. Faridul Hasan, Md. Nahid Pervez, Md. Eman Talukder,
Sakil Mahmud, Vincenzo Naddeo, and Yingjie Cai

Abstract Consumers have constant demands for high-quality handmade items, which has drawn the attention of creative manufacturers and accelerated a significant socio-economic development to the handloom community. Handloom is a symbol of a traditional cultural heritage enriched with distinctive features that have generated a significant flow of earnings to the ancient people of some Southeast Asian countries like Bangladesh, India, Sri Lanka, Nepal, and so on. However, handloom production has been limited to some local areas and families as it requires inherent talents and expertise. However, powerloom technology has brought a revolutionary change to the weaving manufacturing sector in terms of overconsuming resources, generating waste, and increasing production cycles with a critical risk of environmental pollution. Thus, global handloom and textile industries are facing challenges with respect to the sustainability of the technological expansion and larger production volumes that must occur to meet increasing consumer demand. Thus, a new market is developing for ethical and sustainable handloom products. However, in terms of decentralized distribution, the handloom sector contributes to employment creation, financial security, skill development, and cultural integration (identity and diversity). A fair trade and business could build the pillars of sustainability in terms

K. M. F. Hasan · Md. N. Pervez · Y. Cai (✉)
Hubei Provincial Engineering Laboratory for Clean Production and High Value Utilization of Bio-Based Textile Materials, Wuhan Textile University, Wuhan 430200, China
e-mail: yingjiecai@wtu.edu.cn

K. M. F. Hasan
Faculty of Engineering, Simonyi Károly, University of Sopron, Sopron 9400, Hungary

Md. N. Pervez · V. Naddeo
Sanitary Environmental Engineering Division (SEED), Department of Civil Engineering, University of Salerno, via Giovanni Paolo II 132, 84084 Fisciano, SA, Italy

Md. E. Talukder
Shenzhen Institute of Advanced Technology, Chinese Academy of Sciences, Shenzhen 518000, China

S. Mahmud
Ningbo Institute of Material Technology and Engineering, Chinese Academy of Sciences, Ningbo 315201, China

© The Author(s), under exclusive license to Springer Nature Singapore Pte Ltd. 2021
M. Á Gardetti and S. S. Muthu (eds.), *Handloom Sustainability and Culture*,
Sustainable Textiles: Production, Processing, Manufacturing & Chemistry,
https://doi.org/10.1007/978-981-16-5967-6_2

of economic, social, and environmental for this handloom sector. On the other hand, the inclusion of innovative design and technology, closed-loop trade strategy, eco-friendly manufacturing processes, and community-based entrepreneurship could be driving forces for sector expansion. This chapter highlights and demonstrates the different aspects of environmental sustainability in the craft and fashion industries for associated handloom products.

Keywords Handlooms industry · Innovative design · Sustainable technology · Community-based entrepreneurship · Zero wastes · Competitive business

1 Introduction

The world has now realized the significance and importance of creative industries to develop the economies within specific communities through mainstream cultural identity utilization. These creative ideas are associated with the nurturing and focus on innovation-based knowledge through "inspiration/motivation, imaginative thinking, innovation, and uniqueness". However, the development of the new product through ancient heritages via cultural significance reflection is also gaining interest. Creative industries can play vital roles in supporting the economic advancement of developing countries through collaborations with world partnership programs by including vulnerable groups in society under sustainable development strategies. The handloom industry is associated with the expertise, skill, and knowledge that has evolved over time to small-scale production volumes [1, 2].

Weavers have been found scattered around this region (like Bangladesh, India, Sri Lanka, Nepal, and so on) forming a weavers community inside the village or as independent weavers. The main weaving products are sarees (womenswear), menswear, bed covers, curtains, accessories, toys, cushion coverings, pillow covers, and so on [3–8]. The rich finishing, attractive aesthetic appearances, superior product quality, and enriched cultural norms add extra value to the products. Although there are many fashion houses and well-known brands established throughout the world, they are facing constant criticism in terms of sustainable products. In this regard, handloom-based products show a potential route for the production of green and environment-friendly products without harming the environment through consumption [9–13].

Recently, the global, fair trade movement has expanded its scope to the handloom sector. Besides, the constantly increasing interest in sustainable and innovative products [9, 14–18] has facilitated sector expansion. Hence, entrepreneurship within the different communities could be encouraged more for the strategic development and improvement in handloom sectors [19, 20]. Specifically, women are in more vulnerable financial situations in developing countries and could be employed in the growing handloom sector to generate household income. Countries such as Bangladesh, India, Sri Lanka, Nepal, and so on could create indirect and direct employment opportunities and reduce poverty, especially for women, through the

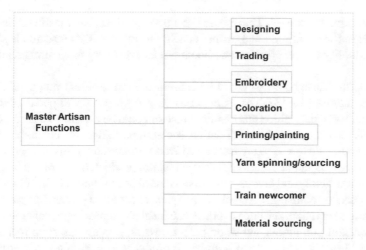

Fig. 1 Responsibilities and tasks performed by a master artisan

promotion of local production [21–24]. This sector is bringing significant contributions to both the local and international marketplace through enhanced economic growth.

Although the traditional handloom sector is a substantial cottage industry in some Southeast Asian nations, it is still suffering from a lack of technological advancement. Production is still extensively dependent on traditional manual labor compared to the modern automated machinery-based manufacturing houses, which significantly affects productivity [25–27]. Expert artisans are key contributors to attractive and fashionable handloom products with wide variation in items through personalized skills, dedication, concentration, and efforts. Some of the key functions of expert artisans are summarized in Fig. 1. The expert craftsmen are capable of producing coarser to finer handloom products from cotton, silk, wool, and so on with plain or other derivatives of woven fabric by using a loom or in collaboration with printing, dyeing, painting, and embroidery experts [26]. The handloom is nearly eight times slower than the powerloom in terms of efficiency [26], and so artisans are competitively challenged against modern textile production based on price and production output/efficiency. The handloom weavers in the United States also face the transition from traditional handloom to powerloom, but as the fancy textiles are still limited with powerlooms, they are still surviving through mechanizing the weaving process [28]. Conversely, the coloration of yarn/fabrics in the handloom sector also appeals to newer approaches like coloration with biosynthesized nanoparticles [29–33]. Besides, nano-based functionalization can also provide handloom fabrics with improved thermal, mechanical, and antibacterial properties, as they are stored for long periods of time by consumers, and this technology could add extended product value [34–36]. Recently, the muslin (the most famous handloom cloth which was extinct long times ago) has been tried to recover the technology and used materials again [37]. A group of researchers from Bangladesh has tried for several years and finally became successful. The research group has produced six muslin fabrics again, identical to the traditional and extinct muslin fabric [37]. This research is showing the potentiality of this precious handloom product recovery and commercialization again.

The term "sustainability" could be expressed from different points of view for different sectors [38–43]. The expressions vary from person to person. However, sustainability in the fashion and handloom sector could be explained in terms of environmental protection, cultural validation, social impartiality, and economic equality [44], although there are many issues involved with sustainability aspects, such as which factors should be followed for which sector, product, or region. In this regard, lots of contradictory and complex information have been found. The handloom sector faces bitter environmental and social consequences for rapid production cycles along with waste generation, resource overuse, unethical and unhealthy labor employment, environmental pollutions, and so on [45]. However, it is necessary to focus on the sustainable production of handloom-based products to enable a more prosperous sector. Besides, the fair business principle and efficient manufacturing approach could minimize sustainability-related criticisms. On the other hand, water sources are becoming polluted near handloom production areas with wastewater discharge [46–48]. Hence, the proper treatment of effluents besides using green dyestuffs is important to reduce environmental burdens.

However, with the rapid development and expansion of science and technology, small-scale productions are turning to large volumes of production. Besides, the responsibility and awareness toward a sustainable product are increasing in terms of environmental perspective. Therefore, it is necessary to bring efficient and more sustainable technology (either in mechanization or chemical treatment) in this sector to compete with the modern textile industries. Proper artisan training for skills, expertise, and knowledge development could also assist in the enhanced development of this sector. In addition, support from different government and non-government organizations could play a big role in developing community-based entrepreneurship among handloom weavers. Furthermore, the background of the handloom sector, manufacturing technology, marketing perspective, employment generation, environmental and coloration aspects, and SWOT analysis would be carried out in this current chapter.

2   Historical Background of Handlooms

The earliest records for cotton weaving were found in 1500 BCE in Bengal as per *Rig Veda*. Other documentation describing weaving was found in Bengal at 800 BCE (*Asvalayana Srauta sutra*), 300 BCE (Megasthenes *Indika*), and in the first century (Ptolemys Geogrphia) [49]. Muslin was exported via silk road to different regions of the globe like Persia, the Roman Empire, etc. in exchange for gold which was used by higher classes of people in society [50]. There was also some information regarding business between China and Bengal for muslin as per Fa Hian (A Chinese traveler and

Buddhists Monk). Trade began to flourish in the seventh and eighth centuries to the Middle East, which had been mentioned by travelers and scholars during this time. Besides, according to Marco Polo, trade was enhanced because of the availability of cotton in this region [51]. Next, Arab traders were replaced by Portuguese traders at the end of the sixteenth century. Later on, the East India Company established a factory in Dacca and took control of muslin trading. Then, Dutch and French traders became involved in this business in the seventeenth and eighteenth centuries. Next, the East India Company of Britain defeated the then ruler of Bengal (Siraj-Ud-Daula) and led muslin trading for the next century. Unfortunately, with a lack of availability of the finest cotton and an expanded industrial revolution with automated machinery during this time, outstanding handloom products have become extinct over time [51].

Jamdani is an attractive and extremely popular sari found in Bangladesh and India. Jamdani is made using hand-weaving technology and the finest cotton yarn along with embellished silver, gold, or silk yarn flowery motifs woven into the surface. However, the motif threads have become heavier/coarser than the warp and weft yarn used for fabrics embroidery [52]. This technology originated in Dhaka, Bangladesh [53]. Although the muslin was patronized by the richest and most powerful peoples of the society, jamdani was not initially as popular in the beginning. The major differences between jamdani and muslin at that time were that muslin was made of the finest spun cotton, but jamdani was made of pulled thread or old/second-hand clothing (like saree, lungi, and dhoti) [53]. Photographs of muslin and jamdani fabric are depicted in Fig. 2. However, another study has mentioned that the word jamdani came from Persian words, *jam* meaning flower and *dani* meaning container/vase [54]. The decorated motifs (leaves, fruits, flowers, creeps, etc.) were inserted into the fabrics through a weaving process using a manually operated loom [55]. The microclimate weather was suitable for jamdani fabric weaving, which required poetically described words like "running water" or "woven air". The weaving of jamdani required two artisans to work together, and this handloom product (muslin) was expensive, as it is required a long time to make the fabrics from the finest specialty cotton. The price for this product was nearly £56 for a single piece during the time of Naib Nazim of Dacca (currently known as Dhaka) in 1778 [56]. During the periods of the Ottoman and Arabian empires in the seventeenth century, the Bengal region and some parts of India became the global center for the world's silk and cotton-based clothing trade, especially for muslin and jamdani [54, 56].

The handloom weaving-based cottage industry is still popular in Bangladesh, especially in Rupshi (part of Narayangonj district), Demra (part of Dhaka division) [54], Kalshi (part of Dhaka division), and the Tangail district. Currently, the Government of Bangladesh and different non-government organizations provide technical, logistical, and financial support to the artisans involved in the handloom sector to keep it viable while competing against the booming automated textile industry. So, the gradual expansion of the handloom sector has been observed in Bangladesh from with the total number of handloom artisans increasing from 85,478 in 1941 to 505,556 in 2003 [54]. According to the labor foundations of Bangladesh, there are 183,000 handloom units in Bangladesh, where nearly 50% are women are employed as spinners, weavers, dyers, associated artisans, and embroiderers [58]. Handloom

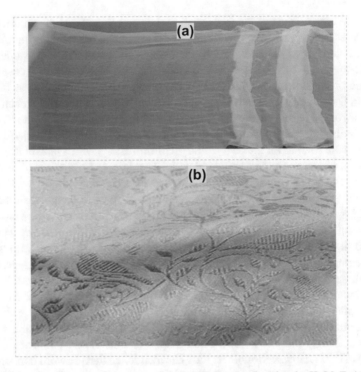

Fig. 2 Ancient handloom products: **a** muslin; digital photographs taken by K. M. Faridul Hasan (courtesy goes to Textile Gallery, National Crafts Museum & Hastakala Academy, New Delhi, India) and **b** Banarasi handloom from pure Katan silk and gold [57]

is a 2000-year-old craft-based technology in Bengal and India. Currently, the handloom industry in India is about 10% of the domestic market share for textile-based items and is a major source of rural livelihood [59]. Handlooms are also found in Sri Lanka with an enriched, craft-based industry employed with many women (which is around 3000 years old technology) [60–62] and also in Nepal [63].

3 Traditional Handlooms

Handlooms are attractive and beautiful fabrics made by artisans. The efficiency depends on the skills of the respective handloom weavers. However, even when using the same design with the same raw materials with the same weavers, the product will most likely be different when made by hand. This is the uniqueness of every handloom product, which has made it quite popular. Handloom fabrics also reflect the mood and skills of every individual weaver as well. Hence, handloom fabrics also become unique pieces of specialty products [64, 65]. There are various types of weaving patterns and styles found based on different regions. Some of the variable styles are tribal motifs, tie-dye, attractive arts on fabrics, simple plain weave fabrics, geometrical designs, colorful combinations of warp and weft yarns, and so on [66–68]. Handloom products (former India but currently, Narayangonj, Bangladesh) were typically exported to China, Egypt, and Rome [66]. At the time, many villages had weavers for producing traditional clothing items like saris, lungis, dhotis, and bed covers. The most interesting thing was that they were created with hand spinning (Charkha wheel used for spinning to make yarn) and weaving [66]. Generally, cotton fibers were cleaned and processed by weavers, but other special fibers like wool, silk, and jute were cultivated and supplied to the artisans by the farmer, shepherds, or foresters.

4 Handloom and Associated Sector

Handlooms were typically made of various types of wooden frames (sometimes also bamboo) used by the artisans for weaving natural-fiber-oriented (like cotton, silk, jute, wool, and so on) woven fabrics [66]. The clothing items were made by expert artisans with/without their family, which is also termed as a cottage industry. Generally, they performed yarn spinning, dyeing, fabric weaving, embroidery, and so on (Fig. 3). However, their produced fabrics are also termed as handloom [66]. The most attractive feature of this technology was that no utilities, like electricity, natural gas, etc., were used for production as the looms were operated by hand. So, this technology is also considered the most environmental-friendly clothing manufacturing method.

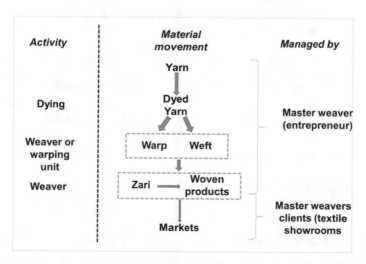

Fig. 3 Value chain and governance of handloom sector. Adapted with permission from Elsevier [70]. Copyright Elsevier, 2015

The handloom sector comprised handicraftsmen and artisans who are experts in producing cottage industry fashionable textiles using hand spinning, handloom-based weaving of fabrics, embellishing them with the manual artistic design using natural coloration and printing and different specialty works like embroidery, bead, mirror, patch, quilting, and so on [69]. Besides, in communities or rural areas, it has also a major source and integral part of earnings, especially for the underprivileged craftsman and the families depending on them (Figs. 3 and 4).

Nowadays, many weaving styles followed by handloom weavers use spun yarns obtained from spinning factories to produce handloom-based fabrics too. In addition, there is still hand-spun yarn too. The fabrics made of hand-spun yarns are termed khadi fabrics [66]. Some of the special handloom products are bed sheets, bed covers, tapestries, lungi, Tangail muslin, checks, jamdani sari (Fig. 6), Benarashi sari/Brocket (Fig. 6), Tangali sari (cotton sari, jamdani sari, soft and half silk sari, balucherri), gamcha, mosquito nets, and sofa covers found in mainstream areas of Bangladesh [71]. However, there are also other specialty handloom products found in tribal areas of Bangladesh like woolen bed sheet, ladies chadar, lungi, bag, and thami (tribal dress) found as special Rakhine wearables in the Patuakhali district and the tribal fashionable wears (khati (orna), thami, lungi, and chadar) found in the Hiltract areas of Chittagong ((Bandarban, Rangamati, and Khagrachari), Miniouri clothing (innachi and orna), three-piece, unstitched clothing, and Munipori sari, etc.) [71]. On the other hand, some of the popular handloom products in India are Daksari,

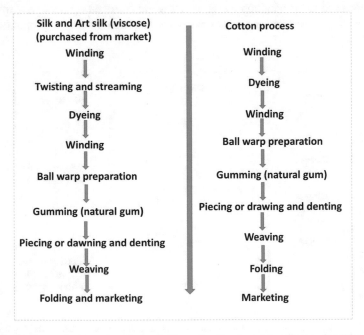

Fig. 4 Production and processing flow of silk and cotton-based handlooms (traditional khana fabric) [76]

kutips, vest coats, chunni, hand bags, purses, mobile covers, kinnauri patti, kinnauri muffler, kinnauri gaachi and dudu, file covers, mats, carpets, cushion covers, slippers, sari, lungi, and so on [64, 72–74]. Besides, coarse clothing, bags, fishing nets, sacks, naturally colored "Allo"/bhanga cloth, vests, head-straps, cushion covers, satchels, and shirts are reported as traditional handloom products in Nepal [63, 75]. A manufacturing flow process of Allo fabric is shown in Fig. 5. The handloom products found in Sri Lanka are laptop covers, soft toys, table mats, key bags, curtains, saris, menswear, womenswear, and so on [45, 60].

Initially, woven fabrics were made of cotton along with simple borders. But, over time, as the cotton-based products became more familiar, the fabrics turned into more attractive products from silk, wool, gold, and so on [52]. Previously, muslin was considered the finest sari which was then made by handloom weavers. Muslin was particularly found in the monsoon landscape of Bengal, which is currently known as Dhaka, and in the deltas of Ganges, Brahmaputra, and Meghna [49]. Muslin has drawn significant attraction in the international market for its extreme transparency, delicateness, and fineness [49]. Although muslin has since disappeared, jamdani (made of cotton, gold, wool, and silk) is becoming another benchmark of handloom-based product that remains popular with consumers.

However, handlooming is a traditional profession enriched with superior cultural values and diversification, providing livelihood to a significant number of people. This sector is reliant on the expertise of artisans who weave by hand. Different fashion houses, retailers, and non-profit organizations are extending their hands to support them (weavers) to continue their operations to survive the influx of new technology [80]. The fabrics made through hand crafting by artisans and small community-based shops provide unique features and cultural significance worldwide. Generally, as the craftsmen are involved with the manual (hand use) operation

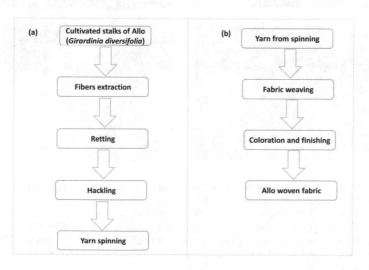

Fig. 5 Production flow process of "Allo" handloom in Nepal

Fig. 6 Different handloom products: **a** Dhakai jamdani sari; **b** Dhakai jamdani sari; **c** Manipuri sari; **d** Half-silk jamdani sari made from 84 finer count yarn on white color; **e** Jamdani sari with embroidery using multi-colored yarn on white surface; **f** Red-colored bridal jamdani from finest yarn. Digital photographs of (**a**), (**b**), (**c**), (**d**), (**e**), and (**f**) were taken by Wali Uddin (Artisan at Rupgonj, Narayanganj, Bangladesh), (photographs collected by Bangladeshi Sari (brand)) [77]; g Traditional jamdani sari found in Mirpur Benarasi palli, Dhaka, Bangladesh [78], and h Traditional bridal/ceremonial clothings [79]

to produce the fabrics, the products become more attractive, decorative, gorgeous, fine, and a symbol of cultural integrity. Currently, social inventions are considered as the development and creation of new ideas for problem-solving (products, models services, processes, and mode of laws, regulations, and provisions) from different socio-structural perspectives [81].

However, this sector is disappearing from so many regions with the expansion of high-tech industries because the hand-crafted products produced by artisans are labor-intensive and time-consuming [82–84], which leads to higher costs than typical textile products made by the automated textile manufacturing companies. Hence, consumers frequently choose low-cost textile goods from markets rather than hand-loom products. So, the artisans are getting forced to sell their products at cheaper prices which are not enough to cover their raw materials or production costs. So, this sector is going to be endangered as the craftsmen are switching to other sectors with better facilities and wages.

5 Technology (Weaving and Coloration) of Handloom

The handloom used by the artisan is a very simple machine with vertical looms, where the harness is fixed (Fig. 7) [71]. A yarn with a specific length and width is processed with a wooden frame through winding [85]. Then the weaving starts with fitting the yarns into heddles of vertical wooden frames of the loom. The main parts of a loom are shown in Fig. 7. Besides, the coloration of handloom products is another significant technology adopted by traditional artisans. The widely used colors for handloom-based products are white, black, green, red, and yellow [85]. As

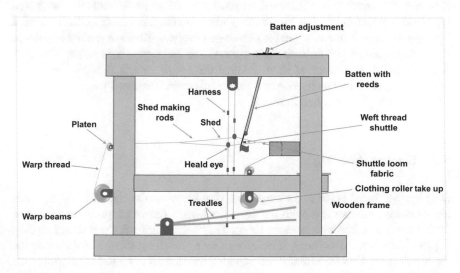

Fig. 7 Schematic diagram of a foot-treadle traditional handloom

Fig. 8 Artisans working with handloom product manufacturing **a** [87], **b** [88], **c** [89], and **d** [90]

the colorants are totally derived from natural sources, they are sustainable products, free of chemical toxicity.

Artisans living in Manipur, India use the handloom as daily household furniture, which is easily movable/shiftable. However, they have modified the traditional loom, but the base is the same and made of bamboo and wood, except the heddles [86]. Smaller plastic pools were made from pencil/ball pen (waste materials) to design the extra weft. Artisans first design the handloom products on normal/tracing paper using kerosene, which is then placed between the lengthwise yarns (warp). However, the quantity of additional crosswise (weft) yarns depends on the design types and woven fabrics. The lengthwise and crosswise yarns and fabrics are treated by applying detergent and neem for bleaching purposes. Khadi, a special handloom fabric manufactured by artisans, uses yarns (handmade) produced using "Charka". A schematic flow process of khadi fabric from cotton or silk is shown in Fig. 4. Sizing was performed using rice and flower of maida. Different weaving activities of artisans are photographed in Fig. 8.

5.1 Carpet Weaving

Initially, the "taan" is made for carpet weaving based on the size of the yarn. Then the weaving process starts with the design, which is drawn on paper (graph). The woven wool is beaten by a hammer (termed as "flag") during the weaving. Labeling is provided when the weaving is completed by a "levelling scissor" for making the carpets thinner. This technology was reported by Sharma et al. for handlooms used in Sikkim, India [85].

Coloration has a very deep history in India. The famous "calico printing" was found in India in the fifteenth century when cotton fabrics were colored by Indigo natural dyes [91]. Coloration is an ancient technique, which was also practiced during the Bronze age times in Europe. Ancient coloration was performed by sticking the plants onto the clothes and rubbing the crushed pigments onto the fabric surface. However, over time, the coloration process has become more sophisticated and the dyestuffs were extracted from plants, fruits, and leaves, which were then incorporated into the fabric through boiling and providing enhanced wash and light fastness properties [92]. However, the fastness of the naturally extracted dyes (Fig. 9) was also good. This technology was reported by Pandya et al. for handloom-based technology at Manipur, India [86]. However, until the nineteenth century, people were mostly using natural colorants (Fig. 10) that were totally safe for the environment

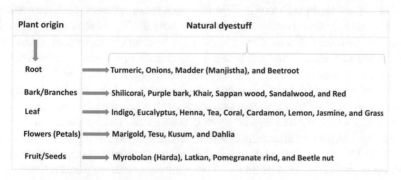

Fig. 9 Different natural dyestuffs extracted from biomass resources

Fig. 10 Photographs of extracted natural colorants from biomass resources. Digital photographs were taken by K. M. Faridul Hasan (courtesy goes to Textile Gallery, National Crafts Museum and Hastakala Academy, New Delhi, India)

[93]. Since then, synthetic colorants have been used for many variations of hues and chroma. Different types of natural colorants (extracted from vegetables, wastage material, and tannin) are still used for producing handloom fabric [94].

6 Community Entrepreneurship of Handloom Sector

Entrepreneurship is considered a way/route for the transformation of resources/raw materials into valuable and useful products or services. As entrepreneurship plays diversified roles in developing the economy, it has extended its scope to people living in different communities as well [95]. Community-based entrepreneurship is a traditional base for the development of economic progress. However, community-based entrepreneurship is a totally different economical approach than the traditional industrial base. It is a profitable way to earn a daily livelihood in small communities especially in rural areas in developing countries. The handloom sector is contributing to improved socio-economic situations in these underprivileged communities [96]. Most of the women in SAARC (South Asian Association of Regional Cooperation) regions had not been directly involved in the direct earnings for the family outside of household work [97, 98]. However, the handloom has given them some opportunities to contribute to family earnings in some communities. The handloom sector has a very strong positioning after agriculture in terms of rural employment creations in India [95]. The major challenges of this sector are unstable markets, an unskilled and uneducated workforce, lack of financial investment, lack of technical advancements, and so on. Although, many in the workforce involved with this sector have basic skills, but there is still a lack of skills in the newly joined workforce. So, they may appeal for special requirements in training or education to improve their skills to compete with other sectors in society equipped with state-of-the-art technology. However, like other SAARC nations, this handloom sector is rapidly shrinking based on critical obstacles and issues. As the people in rural areas have a lack of employment opportunities, financial, training, and technical support in this sector could enhance the employability of folks from underprivileged communities. Moreover, these kinds of initiatives would also motivate them further to not only consume their handloom-based products for personal usage but also sell them into national or international markets, where traditionally these products had been used for personal use only. Hence, more entrepreneurs would be created in the communities. Hereafter, the governments of these regions have helped to eliminate such obstacles and barriers to further extend this sector. Some of the key barriers of weavers are mentioned below:

- Artisans working in handloom sectors are not directly entrepreneurial. In most cases, they depend on the middlemen to gain access to the market, which hinders their potential to get enough payment for their work and efforts toward the products.

- Artisans are not involved in the research and development for innovative designs and products to meet the constantly increasing demand of consumers through competitive fashionable and functional products in terms of powerloom-based textile industries.
- Lack of motivation of younger generations to choose the handloom sector as a prominent and viable career path.
- Lack of technological advancement is not allowing the handloom sector to progress in comparison with the powerloom.

In the case of community-based entrepreneurship in the handloom sector, all members of the community work together to achieve their desired goals. In this way, they also develop their skills collectively and simultaneously which facilitate the function of the entrepreneur [99]. The fundamental purpose of community-based entrepreneurship is groupwise working, mutual trust, respect, and understanding along with the motivation toward achieving their goals [100, 101]. However, it is necessary to link some activities (operational, efficient leadership, strategic problem solving, supporting the development of business endeavor) to promote community-based entrepreneurship in the handloom sector. However, individual entrepreneurship in the community aims at uniqueness in terms of their resources, multiple skills, economic, and social goals [102]. However, in the case of community-based entrepreneurship in rural areas, limitations and obstacles exist in terms of networking to expand the business [103]. Although the journey of the handloom sector has not always been easy, one suggestion for "globalization" is the collaborative model for progress and development through the empowerment of local companies/organizations with modern equipment, technology, and scientific advancements.

7 Sustainable and Fair Trade Approach

The fair labor and green movement is gaining momentum throughout the world. The consumers of products are demanding products that are made ethically through ensuring environmental sustainability. This shift of the consumer's purchasing approach and behavior is turning manufacturers to be more sustainable. So, the weavers of the handloom sectors are gaining an ethical working environment along with standard wages, depending on their locations. This scenario also helps to alleviate the poverty conditions of economically and socially marginalized weavers/peoples engaged in the handloom sector. In addition, it is important to set up marketing strategies for handloom products to ensure efficient trading output. In this regard, 4P (product, price, promotion, and place), which is also the pillars of a marketing strategy (Fig. 11) could play a vital role through market mix investigation. It could maximize the possibilities to achieve more customers and business.

Fig. 11 4Ps related to
handloom trading

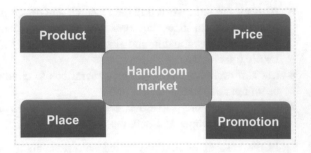

The term "fair trade" also represents the ethical approach toward product manufacturing [104–106]. In addition, reducing environmental impact is a key objective of fair trade. The generation of production-based waste is becoming a growing concern/challenge worldwide. Besides, waste generated from production houses is considered inefficient during manufacturing. On the other hand, it also impacts the monetary flow for the use of excessive raw materials, which are then discarded without efficient utilization in the final stages [45]. It has been prioritized to minimize waste generation through sustainable production technology implementation [45, 107, 108]. Also, if waste generation could be minimized in the handloom sector, then the production volume using the same raw materials would be improved. If the total output is increased, then the associated production costs decrease per unit and revenue will increase, as shown in Fig. 12.

On the other hand, a zero-waste strategic concept [109–111] also focuses on designing the production process in a way that could facilitate avoiding any kind of extra waste generation. This strategy could also save the environment by reducing the burdens of discharged pollutants to the environment. This is in conjunction with the textile and associated industries using state-of-the-art technologies to convert this waste into useful, cost-effective, efficient, and beneficial products [112–116]. However, the handloom sector could facilitate economic development to the local

Fig. 12 Costs and revenues
versus average fixed costs

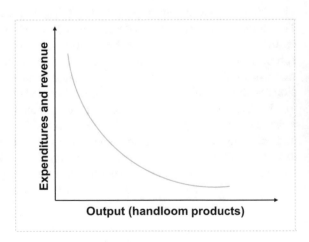

community through implementing sustainable business strategies and ethical trade promotions. Specifically, women and low caste peoples are direct beneficiaries of the handloom sector in developing nations.

8 Lack of Knowledge, Training, and Education Within Craftsman

This is the era of technological advancements, digitalization, and globalization. The mode of rapid, internet-based communication has made the online market extremely popular with manufacturers and consumers. However, artisans involved in handloom sectors are still deprived of internet-based information and communication because of their lack of educational knowledge [117]. Most often craftsmen cannot meet with the customer directly, and they are unaware of consumer expectations on product design and cost. So, they are unable to achieve high enough prices as they need to depend on middlemen for trading and marketing purposes. Also, artisans are unable to learn about scientifically advanced technology to create more automation to enhance the productivity and quality of products due to a lack of knowledge [118]. Conversely, it also becomes challenging to collect better raw materials (yarn, accessories, colorants, and so on), both from home and abroad with competitive prices and quality. So, many handloom products are unable to be sold in widespread international markets. Likewise, this lack of knowledge leaves a gap between the latest trends, design, and the expectation of worldwide customers' demands and requirements and what the artisan knows [119]. Consequently, community-based entrepreneurship could not compete efficiently with well-equipped automated textile manufacturing companies in terms of productivity and costs. Besides, there are enough skill-based problems with the new weavers in the handloom sectors. Therefore, adequate training for people regarding handlooms could motivate the craftsman extensively and improve their skills to innovate better products.

9 Environmental Aspects of Handloom

The demand for clothing from the handloom has increased throughout the world in terms of constantly rising environmental concerns. Handloom products are not dependent on strongly mechanized production houses like other industrial manufacturing units and save energy [120, 121]. This sector is also gaining attention for the reduced negative impacts on the surrounding ecological environment. However, handlooms are also generating significant amounts of solid and liquid waste, which are major effluents into the environment. Handloom weaving creates both hard and soft wastes [122, 123]. Soft wastes are generated by handlooms before the spinning process. However, the wasted materials after the spinning and twisting processes are termed

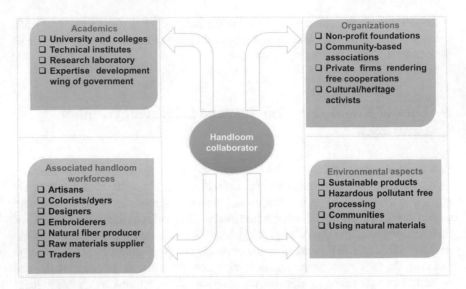

Fig. 13 Different collaborator/stakeholders involved in handloom sector

as hard waste. The wasted materials created from handlooms include the wastes from soft fibers, hard fiber (wastes from yarn spinning), beaming wastes, spool, creel, off-cut, and packaging wastes [122]. Conversely, elimination/proper treatment of poisonous pollutants from the effluents of yarn processing through applying the traditional coloration process is becoming an alarming concern to artisans (Fig. 13).

Handloom materials (like yarn, fiber, or fabric) require coloration, and the coloration process needs huge amounts of water. Besides, natural and synthetic dyestuffs and chemicals used for dyeing are responsible for enhanced biochemical oxygen demand (BOD), chemical oxygen demand (COD), dissolved oxygen (DO), and deviated pH from standards. Besides, there are other risks associated with changing color, composition, and odor by the discharged effluents for both surface water and groundwater [47]. Some of the studies have reported that the effluents discharged from the handloom industry of Ranaghat (West Bengal, India), which has a big cluster of handloomers near the river Churni, accumulate chromium (Cr) in water [124]. They also found the presence of Cr in aquatic plants, mollusks, and weeds [124]. Another study has revealed that handloom effluents from the Madhabdi area of Bangladesh contain heavy metals (iron 3.81, chromium 1.35, copper 1.7, lead 0.17, manganese 0.75, and zinc 0.73 mg/L) [47]. They also mentioned that the metal contents exceed the standard limits for discharged effluents [47]. The treatment of wastewater discharged from handloom industries could be treated through several methods like biological, physical, and oxidation approaches (Fig. 14a). A schematic layout plan is also provided in Fig. 14b for effluents treatment discharged from manufacturing units.

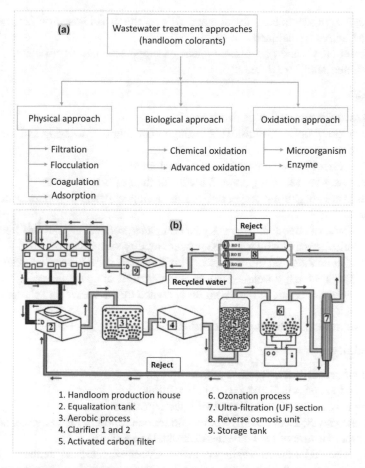

Fig. 14 Wastewater treatment: **a** treatment methods for discharged colorants and **b** processing of discharged waste water from production houses. Adapted with permission from Elsevier [125, 126]. Copyright Elsevier, 2016 (**a**) and 2020 (**b**)

10 SWOT Analysis of the Handloom Sector

Strength

- Skilled artisans/handloom weaver's availability
- Family business supported by all family members
- Preservation of traditional production methods and cultural heritage
- As the handloom business is run in home/rural areas, the necessity for additional house/space rent is not required in most cases

- Handloom products have the demand all over the world based on fancy designs and attractive appearances
- Different NGOs and government agencies support the handloom sector
- Developed retail structure.

Weakness

- This sector is not well-organized
- Generation of pollutants/effluents through fiber/fabric processing and associated colorations
- Lack of direct contact/interaction with customers
- Artisans are not receiving enough credit for their efforts and work
- Artisans are not gaining proper remuneration/financial benefit with respect to their labor
- Due to lack of adequate facilities, the complete products are not finished by the same units/artisans, which creates product quality variability
- Lack of education and internet access hinders their communication externally to know about the latest trends, fashion, and demands of consumers
- No showrooms outside their community owned by the weavers to showcase their product items
- Lack of modern technological advancements.

Opportunity

- Diversification of handloom products
- Tremendous popularity of the designed outfits
- Artisans could be directly linked with the consumers
- Artisans and their community could read market reports or gain market access to make them aware of the latest trends around the globe
- Designs could be modified/altered per consumer expectations/trends
- Arranging fairs and trade shows to exhibit handloom products to consumers
- Developing the brands of artisans like jamdani/muslin
- Possibility of promoting e-commerce-based business strategies.

Threats

- The world is shifting to automated industrial manufacturing units, which is a big challenge for the manually operated handloom sector
- Powerlooms with enhanced efficiency and production volume is threatening the handloom segment
- Artisans are not encouraged to involve the next generation in this sector
- Weavers are becoming demotivated for comparatively less salary/remunerations compared to other textile industry employees
- Lack of logistical support
- Inadequate functional properties of the handloom products beyond aesthetics.

11 Conclusions

Handloom products are heavily associated with traditional, historical, religious, identity-based cultural values of the community. Bangladesh, India, Nepal, and Sri Lanka have had glorious pasts in handloom-based products. However, artisans/handloom weavers are still considered as lower class people in society based on lower/no education, low wages, absence of better living conditions, inadequate intake of nutrients, and unusual lifestyles. Hence, it is also necessary to improve the education, skill, and lifestyles of the artisans to make this handloom sector more sustainable, innovative, and potential. However, artisans also need to focus more on innovations in design, fabrication, comfort, and attractive handloom products. State-of-the-art manufacturing technology with enhanced handloom product quality is also another important aspect that needs attention. The "zero waste" and lean manufacturing in the handloom sector could maximize the efficient utilization of resources and increase productivity. Furthermore, effluents discharged from handloom production houses need adequate treatment to ensure no harmful effects on the environment. Conversely, handloom weavers need to enhance productivity through waste minimization to reduce the product price compared to other textile wearables to gain consumer interest. Besides, community-based management of the handloom sector could boost the handloom business through collective and diverse efforts.

References

1. Broudy E (1993) The book of looms: a history of the handloom from ancient times to the present. University Press of New England, New York, USA
2. Beddig C (2008) Cluster development policy rooted in the collective efficiency approach: an effective poverty alleviation tool in the Indian handloom sector? Graduate Institute of International and Development Studies, Geneva
3. Anthony AA, Joseph MT (2014) SMEs in Indian textiles. Palgrave Macmillan, London, United Kingdom. https://doi.org/10.1057/9781137444578
4. Singh R (2010) The fabric of our lives: the story of Fabindia. Penguin groups, New York, United States
5. Tod O (2012) The joy of handweaving. Douver Publications Inc., New York, United States
6. Matthews K (2018) Encyclopaedic dictionary of textile terms, vol 3. Woodhead Publishing India Pvt. Ltd., New Delhi, India
7. Mitra A, Majumdar A, Ghosh A, Majumdar PK, Bannerjee D (2015) Selection of handloom fabrics for summer clothing using multi-criteria decision making techniques. J Nat Fibers 12(1):61–71
8. Anumala K (2015) A demographic study on customer satisfaction about handloom products: a study on Andhra Pradesh state handloom weavers' cooperative society limited (APCO) showrooms. J Text Appar Technol Manag 9(3):1–16
9. Gopalakrishnan D (2021) Reviews on perception of sustainable green consumption practices and its impact on greener lifestyles. Department of Fashion Technology, PSG College of Technology, Coimbatore. https://d1wqtxts1xzle7.cloudfront.net/54308018/GREEN_APP ARELS.PDF?1504245549=&response-content-disposition=inline%3B+filename%3DREVI EWS_ON_PERCEPTION_OF_SUSTAINABLE_GRE.pdf&Expires=1610532130&Signat ure=WGQhpRoMKPz9SeNgAMxcaXtqq5ZmvH1wOoQdajLegABtl3BoHcXfNmdBpOiG

brvq5~v3Cz8k3e3SSBd3UCLuO5T9ou2k26Aiofh4Avh2zYOMJBzzjBSnYWdK4w~GLr
9VnyrUzlxGV~lRutR4s1AyP757iWvZndxnzUBp9NVCQUgq03CsPCvg15bELWs945Qb
qAHIkLPBoyQPSyobn4LVzf1bW6AjPRVSBXtbV-iWnW0r6DlGhmjzvbygzdixo6fy8DYn
LTjQVP8S5MKHkHHpYKuND6skayddONG2BdMA02jxjka793v76RG9u~s9muTTBAf
IEF0kzXRiZ9CLbXOc9A__&Key-Pair-Id=APKAJLOHF5GGSLRBV4ZA. Last Accessed
13 Jan 2021

10. Paulose V, Jayalakshmi I (2018) Analysis of anti-bacterial efficacy in bamboo and bamboo/cotton handloom fabrics finished with selected herbal extracts. Int J Appl Soc Sci 5(10):1626–1633
11. Wani SA, Sofi AH (2018) Green technology for Pashmina Shawl manufacturing. J Fibre Finish 57:43–46
12. Bhalla K, Kumar T, Rangaswamy J, Siva R, Mishra V (2020) Life cycle assessment of traditional handloom silk as against power-loom silks: a comparison of socio-economic and environmental impacts. In: Green buildings and sustainable engineering. Springer, pp 283–294
13. Vernekar S, Venkatasubramanian K (2013) Green initiatives in indian industry: a silhouette. J Commer Manag Thought 4(4):822–836
14. Kane F, Shen J, Morgan L, Prajapati C, Tyrer J, Smith E (2020) Innovative technologies for sustainable textile coloration, patterning, and surface effects. In: Sustainability in the textile and apparel industries. Springer, pp 99–127
15. Pandit P, Singha K, Shrivastava S, Ahmed S (2020) Overview on recycling from waste in fashion and textiles: a sustainable and circular economic approach. In: Recycling from waste in fashion and textiles: a sustainable and circular economic approach. Scrivener Publishing LLC, Beverly, MA, USA1
16. Neto GCdO, Correia JMF, Silva PC, Sanches AGdO, Lucato WC (2019) Cleaner production in the textile industry and its relationship to sustainable development goals. J Clean Prod 228:1514–1525
17. Sharma A, Bhagat S, Suri M (2020) An approach to sustainable livelihood. Social Entrepreneurship and Sustainable Development, Routledge, Taylor & Francis, New York, USA
18. Satapathy SK, Kanungo S (2018) Environment friendly industrial growth for sustainability. Int J Life Sci Earth Sci 1(1):1–14
19. Goswami K, Hazarika B, Handique K (2019) Socio-cultural motivation in women's entrepreneurship: exploring the handloom industry in Assam. Asian J Women's Stud 25(3):317–351
20. Hazarika B (2017) Decomposition of gender income gap in rural informal micro-enterprises: an unconditional quantile approach in the handloom industry, vol 216. National Institute of Public Finance and Policy, New Delhi, India
21. Hazarika B, Goswami K (2016) Do home-based micro-entrepreneurial earnings empower rural women? Evidence from the handloom sector in Assam. Asian J Women's Stud 22(3):289–317
22. Rahman MA, Noman SMH (2019) Poverty and food security analysis of handloom weaver households in a selected area of Bangladesh. J Bangladesh Agric Univ 17(1):80–85
23. Safna N, Nufile A (2016) Women empowerment through weaving industry: a case study in Maruthamunai, Sri Lanka. In: 6th international symposium. South Eastern University, Sri Lanka, pp 658–665
24. Hasan K, Mia MS, Rahman M, Ullah A, Ullah M (2016) Role of textile and clothing industries in the growth and development of trade & business strategies of Bangladesh in the global economy. Int J Text Sci 5(3):39–48
25. John KR, Kamini S (2016) Socio economic status of women entrepreneurs in handloom sector. Int J Appl Home Sci 3(12):459–469
26. Nazari A (2019) Superior self-cleaning and antimicrobial properties on cotton fabrics using nano titanium dioxide along with green walnut shell dye. Fibers Polym 20(12):2503–2509
27. Hasan K (2015) Study on the changes of gsm (gm/m^2) of grey knitted fabric from pretreatment to finishing. Int J Text Sci 4(6):119–136

28. Anderson CS, Spivak SM (1990) A re-examination of American antebellum handloom technology for weaving fancy coverlets and carpets. Text Hist 21(2):181–201
29. Mahmud S, Pervez MN, Hasan KF, Abu Taher M, Liu H-H (2019) In situ synthesis of green AgNPs on ramie fabric with functional and catalytic properties. Emerg Mater Res 1–11. https://doi.org/10.1680/jemmr.19.00012
30. Hasan KF, Wang H, Mahmud S, Taher MA, Genyang C (2020) Wool functionalization through AgNPs: coloration, antibacterial, and wastewater treatment. Surf Innov 9(1):25–36. https://doi.org/10.1680/jsuin.20.00031
31. Hasan KF, Wang H, Mahmud S, Genyang C (2020) Coloration of aramid fabric via in-situ biosynthesis of silver nanoparticles with enhanced antibacterial effect. Inorg Chem Commun 119:1–8. https://doi.org/10.1016/j.inoche.2020.108115
32. Hasan KMF, Wang H, Mahmud S, Jahid MA, Islam M, Jin W, Genyang C (2020) Colorful and antibacterial nylon fabric via in-situ biosynthesis of chitosan mediated nanosilver. J Market Res 9(6):16135–16145. https://doi.org/10.1016/j.jmrt.2020.11.056
33. Hasan KF, Horváth PG, Horváth A, Alpár T (2021) Coloration of woven glass fabric using biosynthesized silver nanoparticles from Fraxinus excelsior tree flower. Inorg Chem Commun 126:108477. https://doi.org/10.1016/j.inoche.2021.108477
34. Mahmud S, Pervez N, Taher MA, Mohiuddin K, Liu H-H (2020) Multifunctional organic cotton fabric based on silver nanoparticles green synthesized from sodium alginate. Text Res J 90 (11–12):1224–1236. https://doi.org/10.1177/0040517519887532
35. Hasan KMF, Horváth PG, Alpár T (2020) Potential natural fiber polymeric nanobiocomposites: a review. Polymers 12(5):1072. https://doi.org/10.3390/polym12051072
36. Ahmad I, Kan C-w (2017) Visible-light-driven, dye-sensitized TiO_2 photo-catalyst for self-cleaning cotton fabrics. Coatings 7(11):1–13
37. Azad AKM (2021) Muslin belongs to Bangladesh. The daily Prothom Alo. https://en.prothomalo.com/bangladesh/good-day-bangladesh/rat-farm-in-rajshahi-strikes-success. Last Accessed 3 Jan 2021
38. Hasan KMF, Péter GH, Gábor M, Tibor A (2021) Thermo-mechanical characteristics of flax woven fabric reinforced PLA and PP biocomposites. Green Mater 1–9. https://doi.org/10.1680/jgrma.20.00052
39. Hasan KMF, Péter György H, Tibor A (2020) Thermomechanical behavior of methylene diphenyl diisocyanate-bonded flax/glass woven fabric reinforced laminated composites. ACS Omega 6(9):6124–6133. https://doi.org/10.1021/acsomega.0c04798
40. Alexander TH, Sakshi GA, Manisha G Quality charaterstics of recycled handloom fabrics. Pharma Innov J 9(3):350–352
41. Muthu SS (2017) Evaluation of sustainability in textile industry. In: Sustainability in the textile industry. Springer, Singapore. https://doi.org/10.1007/978-981-10-2639-3
42. Hasan KMF, Horváth PG, Alpár T (2021) Lignocellulosic fiber cement compatibility: a state of the art review. J Nat fibers. https://doi.org/10.1080/15440478.2021.1875380
43. Zhou S, Zeng H, Qin L, Zhou Y, Hasan KMF, Wu Y (2021) Screening of enzyme-producing strains from traditional Guizhou condiment. Biotechnology & Biotechnological Equipment
44. Parker E (2011) Steps towards sustainability in fashion: snapshot Bangladesh. In: Ammond L, Wigginson H, Williams D (eds). London College of Fashion London, United Kingdom, pp 1–25
45. Dissanayake D, Perera S, Wanniarachchi T (2017) Sustainable and ethical manufacturing: a case study from handloom industry. Text Cloth Sustain 3(1):2
46. Nilofar Nisha J, Arun Prakash M, Vignesh P, Bharath Ponvel M, Kirubakaran V Renewable energy integrated waste water treatment for handloom dying units: an experimental study. Res J Chem Environ 14(1):66–69
47. Nahar K, Chowdhury MAK, Chowdhury MAH, Rahman A, Mohiuddin K (2018) Heavy metals in handloom-dyeing effluents and their biosorption by agricultural byproducts. Environ Sci Pollut Res 25(8):7954–7967
48. Ratnapandian S (2019) Natural colorants and its recent development. In: Sustainable technologies for fashion and textiles. Woodhead Publishing, Duxford, United Kingdom, pp 189–203

49. Hossain L (2020) A critical reading of the dry and permanent ground through the practice of muslin weaving. Monsoon [+ other] grounds. University of Westminster London, United Kingdom, pp 113–120
50. Schoff W (1912) The periplus of the Erythrean Sea. With annotations. Longmans, Green and co, New York, United States
51. Taylor J (1851) A Descriptive and historical account of the cotton manufacture of Dacca in Bengal. John Mortimer, London
52. Mamidipudi A, Bijker WE (2018) Innovation in Indian handloom weaving. Technol Cult 59(3):509–545
53. Chen MA (1984) Kantha and jamdani: revival in Bangladesh. India Int Cent Q 11(4):45–62
54. Ashmore S (2018) Handcraft as luxury in Bangladesh: weaving jamdani in the twenty-first century. Int J Fash Stud 5(2):389–397
55. Crill R (2003) Textiles from India: the global trade: papers presented at a conference on the Indian textile trade. Seagull Books Pvt Ltd, Kolkata, India
56. Reilly T (2014) Muslin, Sonia Ashmore. Text Cloth and Cult 12(1):132–133. https://doi.org/10.2752/175183514x13916051793794
57. Tilfi (2020) White-gold pure katan silk Banarasi handloom fabric. https://www.tilfi.com/products/white-gold-pure-katan-silk-banarasi-handloom-fabric. Last Accessed 26 Nov 2020
58. Altermann W, Kazmierczak J, Oren A, Wright D (2006) Cyanobacterial calcification and its rock-building potential during 3.5 billion years of Earth history. Geobiology 4(3):147–166
59. Mamidipudi A (2019) Crafting Innovation, weaving sustainability: theorizing Indian handloom weaving as sociotechnology. Comp Stud South Asia Afr Middle East 39(2):241–248
60. Wanniarachchi T, Dissanayake K, Downs C (2020) Improving sustainability and encouraging innovation in traditional craft sectors: the case of the Sri Lankan handloom industry. Res J Text Appar 24(2):1–25. https://doi.org/10.1108/RJTA-09-2019-0041
61. Vimalkumar R (2018) Future and challenges of the handloom industry in Jaffna, Sri Lanka. In: 7th international conference, Sri Lanka, 2018. Sri Lanka Forum of University Economists (SLFUE), pp 188–192
62. Costa Y, Fernando P, Yapa U (2018) The effect of ethnocentrism and patriotism on consumer preference (special reference to handloom products in Sri Lanka). J Manag Tour Res 1(2):1–20
63. Subedee BR, Chaudhary RP, Uprety Y, Dorji T (2018) Socio-ecological perspectives of Himalayan giant nettle (Girardinia diversifolia (link) friis) in Nepal. J Nat Fibers 17(1):9–17. https://doi.org/10.1080/15440478.2018.1458684
64. Sharma A (2009) Unique Handloom Products of Garo Tribe of Meghalaya. J Community Mobilization Sustain Dev 4(1):64–70
65. Roy T (2017) Trade and industry II: Pakistan, Bangladesh, Sri Lanka, and Nepal. In: The economy of South Asia: from 1950 to the present. Springer International Publishing, Cham, pp 215–237. https://doi.org/10.1007/978-3-319-54720-6_9:215-237
66. Trust CA (2020) Handloom a timeless tradition. http://www.chinmayaupahar.in/blog/handloom/. Last Accessed 11 Nov 2020
67. Kaur R, Gupta I (2016) Peacock motif in Phulkari: a comprehensive analysis. Thapp J 242–253
68. Okeke C (1977) Factors which influenced igbo traditional woven designs for apparel fabrics. Text Hist 8(1):116–130
69. Jain M (2018) Challenges for sustainability in textile craft and fashion design. Int J Appl Home Sci 5(2):489–496
70. Bhagavatula S, Elfring T, Van Tilburg A, Van De Bunt GG (2010) How social and human capital influence opportunity recognition and resource mobilization in India's handloom industry. J Bus Ventur 25(3):245–260
71. Rahman MM (2013) Prospects of handloom industries in Pabna, Bangladesh. Glob J Manag Bus Res 13(5):1–11
72. Singh R, Chandra Y (2020) Role of handloom households in rural economy. Yojana November 2020 (English)(special edition): a development monthly, p 56
73. India H (2020) Handloom products in India. http://www.handicraftsindia.org/handlooms/handloom-products-india/. Last Accessed 16 Nov 2020

74. Chaudhury T, Joshy K (2019) Occupational divergence in handloom industry: case study of Howrah district. Adv Econom Bus Manag 6(1):79–82
75. Deokota R, Chhetri R (2009) Traditional knowledge on wild fiber processing of allo in Bhedetar of Sunsari district, Nepal. Kathmandu Univ J Sci Eng Technol 5(1):136–142
76. (Fiber2Fashon.com) F (2020) Guledgudd khana: historical heritage of Indian handloom weaving industry. https://www.fibre2fashion.com/industry-article/7534/guledgudd-khana-historical-heritage-of-indian-handloom-weaving-industry. Last Accessed 26 Nov 2020
77. Uddin W (2020) Bangladeshi sari. https://www.facebook.com/Bangladeshi-Shari-%E0%A6%AC%E0%A6%BE%E0%A6%82%E0%A6%B2%E0%A6%BE%E0%A6%A6%E0%A7%87%E0%A6%B6%E0%A7%80-%E0%A6%B6%E0%A6%BE%E0%A7%9C%E0%A6%BF-108024811113699/. Last Accessed 26 Nov 2020
78. Dreamstime (2020) A traditional Jamdani saree in Mirpur Benarashi Palli Dhaka, Bangladesh. https://www.dreamstime.com/traditional-jamdani-saree-mirpur-benarashi-palli-dhaka-bangladesh-partition-some-families-migrated-to-parbatipur-image104153033. Last Accessed 26 Nov 2020
79. Dreamstime (2020) Beautiful traditional marriage ceremonial saree background photograph. https://www.dreamstime.com/stock-photo-part-traditional-saree-marriage-ceremonial-background-photograph-beautiful-image65919842. Accessed 26th November 2020
80. Osterczuk K (2021) Perspectives for more inclusive and impactful fair trade. Analysis of the handloom industry in Nepal. University of Warsaw, Poland. http://www.wz.uw.edu.pl/portaleFiles/6133-wydawnictwo-/Management_challenges/Perspectives_for_More_Inclusive_and_Impactful.pdf. Last Accessed 13 Jan 2021
81. Chiappero-Martinetti E, Houghton Budd C, Ziegler R (2017) Social innovation and the capability approach—introduction to the special issue. J Hum Dev Capab 18(2):141–147. https://doi.org/10.1080/19452829.2017.1316002
82. Bhalla K, Kumar T, Rangaswamy J (2018) An integrated rural development model based on comprehensive life-cycle assessment (LCA) of khadi-handloom industry in rural India. Procedia CIRP 69:493–498
83. Gurumoorthy T, Rengachari N (2002) Problems of handloom sector. In: Soundarapandian M (ed) Small scale industries: problems of small-scale industries, vol 1. Concept Publishing Company, New Delhi, India, pp 168–178
84. Khatoon R, Das A, Dutta B, Singh P (2014) Study of traditional handloom weaving by the Kom tribe of Manipur. Indian J Tradit Knowl 3(1):596–599
85. Sharma TP, Borthakur S (2010) Traditional handloom and handicrafts of Sikkim. Indian J Tradit Knowl 9(2):375–377. http://hdl.handle.net/123456789/8188
86. Pandya A, Thoudam J (2010) Handloom weaving, the traditional craft of Manipur. Indian J Tradit Knowl 9(4):651–655
87. Anni AZ (2020) Handloom industry: dooming, not booming. Daily sun. https://www.daily-sun.com/post/398978/Handloom-Industry-:-Dooming-Not-Booming. Last Accessed 26 Nov 2020
88. Chain TV (2020) Promotion of traditional textile-making skills. https://textilevaluechain.in/2019/07/12/promotion-of-traditional-textile-making-skills/. Accessed 26th November 2020
89. Federal T (2020) 'Knot' an easy job for textile weave revivalists. https://thefederal.com/the-eighth-column/knot-an-easy-job-for-textile-handloom-weave-revivalists/. Last Accessed 26 Nov 2020
90. Standard TB (2020) Traditional weaving in Pabna at risk as losses drive weavers away. https://tbsnews.net/economy/industry/traditional-weaving-pabna-risk-losses-drive-weavers-away#lg=1&slide=0. Last Accessed 26 Nov 2020
91. Datye KV (1991) The textile coloration industry in India. Rev Prog Color Relat Top 21(1):86–97
92. Belemkar S, Ramachandran M (2015) Recent trends in Indian textile industry-exploring novel natural dye products and resources. Int J Text Eng Process 1(3):33–41
93. New YY, Khaing MM, Thwe KM, Wai TP (2020) Extraction of plants dye from different parts of banda (Terminalia cattapa L.) and its coloration effect on clothes. 3rd Myanmar Korea Conf Res J 3(3):1114–1123

94. Das M, Basak S (2018) Advancement of khadi textile: textile glummer of India. Appl Chem Eng 1(4). https://doi.org/10.24294/ace.v1i4.584
95. Kungwansupaphan C, Leihaothabam JKS (2016) Capital factors and rural women entrepreneurship development. Gend Manag Int J 31(3):207–221. https://doi.org/10.1108/GM-04-2015-0031
96. Wanniarachchi T, Dissanayake D, Downs C (2018) Exploring opportunities and barriers of community based entrepreneurship within handloom communities in Sri Lanka.
97. Kavitha L, Mathiazhagan MA, Kannan A, Sivakumar R Structural transformation in women employment: a study on SAARC nations. J Shanghai Jiaotong Univ 16(7):260–272
98. Hque U (2017) Contribution of women entrpreneurs in SMEs amoung SAARC countries. Int J Human Soc Sci 6(6):41–52
99. Tshikovhi N (2014) Importance of Sustainable Entrepreneurship Development in Kwamhlanga-Moloto Village, South Africa. SSRN Electronic Journal
100. Korsching PF, Allen JC (2004) Locality based entrepreneurship: a strategy for community economic vitality. Community Develop J 39(4):385–400
101. Parwez S (2017) Community-based entrepreneurship: evidences from a retail case study. J Innov Entrep 6(1):1–16
102. Peredo AM, Chrisman JJ (2006) Toward a theory of community-based enterprise. Acad Manag Rev 31(2):309–328
103. Johnstone H, Lionais D (2004) Depleted communities and community business entrepreneurship: revaluing space through place. Entrep Reg Dev 16(3):217–233
104. Baines T, Brown S, Benedettini O, Ball P (2012) Examining green production and its role within the competitive strategy of manufacturers. J Ind Eng Manag (JIEM) 5(1):53–87
105. Doherty B, Smith A, Parker S (2015) Fair trade market creation and marketing in the Global South. Geoforum 67:158–171
106. Khan MI, Khan S, Haleem A (2019) Compensating impact of globalisation through fairtrade practices. In: Globalization and development. Springer, Cham, Switzerland, pp 269–283. https://doi.org/10.1007/978-3-030-11766-5_9:269-283
107. Yalcin-Enis I, Kucukali-Ozturk M, Sezgin H (2019) Risks and management of textile waste. In: Nanoscience and biotechnology for environmental applications. Springer, Cham, Switzerlnd, pp 29–53. https://doi.org/10.1007/978-3-319-97922-9_2:29-53
108. Gupta V, Arora M, Minhas J (2020) Innovating opportunities for fashion brands by using textile waste for better fashion. In: Recycling from waste in fashion and textiles: a sustainable and circular economic approach. John Willey & Sons, Beverley, MA, United States, pp 101–121
109. Senanayake R, Gunasekara Hettiarachchige V (2020) A zero-waste garment construction approach using an indigenous textile weaving craft. Int J Fash Des Technol Educ 13(1):101–109
110. Rathinamoorthy R (2019) Circular fashion. In: Muthu SS (ed) Circular economy in textiles and apparel. Woodhead Publishing, Elsevier, Cambridge, England, pp 13–48
111. Franco-García M-L, Carpio-Aguilar JC, Bressers H (2019) Towards zero waste, circular economy boost: waste to resources. In: Towards zero waste, vol 6. Springer, Cham, switzerland, pp 1–8. https://doi.org/10.1007/978-3-319-92931-6_1:1-8
112. Hynes NRJ, Kumar JS, Kamyab H, Sujana JAJ, Al-Khashman OA, Kuslu Y, Ene A, Suresh B (2020) Modern enabling techniques and adsorbents based dye removal with sustainability concerns in textile industrial sector-A comprehensive review. J Clean Prod 272:1–17. https://doi.org/10.1016/j.jclepro.2020.122636
113. Chauhan G (2016) An analysis of the status of resource flexibility and lean manufacturing in a textile machinery manufacturing company. Int J Organ Anal 24(1):107–122. https://doi.org/10.1108/IJOA-11-2012-0625
114. Bhatia D, Sharma A, Malhotra U (2014) Recycled fibers: an overview. Int J Fiber Text Res 4(4):77–82
115. Balachander K, Amudha A (2019) Energy economy recommendations in textile mill. Int J Eng Adv Technol 8(4):168–176

116. Dwivedi A, Dwivedi A (2013) Role of computer and automation in design and manufacturing for mechanical and textile industries: CAD/CAM. Int J Innov Technol Explor Eng 3(3):8
117. Jain M Challenges for sustainability in textile craft and fashion design
118. Rai SK (2015) Ways to modernisation and adaptation: the state, weaving training schools and handloom weavers in early twentieth century united provinces, India. Indian Hist Rev 42(2):261–287
119. Emmett D Conversations between a foreign designer and traditional textile artisans in India: design collaborations from the artisan's perspective. In: Textile society of America symposium, Los Angels, United states, 10–14th September 2014. Textile Society of America at DigitalCommons@University of Nebraska—Lincoln
120. Mamidipudi A, Bijker W (2012) Mobilising discourses: handloom as sustainable socio-technology. Econ Pol Wkly 47(25):41–51
121. Bhalla K, Kumarí T, Rangaswamy J, Siva R Life cycle assessment of traditional kkS handloom silk as against power-loom silks: a comparison of socio-economic and environmental impacts. In: Green buildings and sustainable engineering. Springer nature, Singapore, pp 283–294
122. Pandit P, Shrivastava S, Maulik SR, Singha K, Kumar L (2020) Challenges and opportunities of waste in handloom textiles. In: Recycling from waste in fashion and textiles: a sustainable and circular economic approach, p 123
123. Arolin E (2019) Salvage the selvedge!: upcycling selvedge waste from industrial weaving, using handweaving techniques. University of Borås, Boras, Sweden
124. Sanyal T, Kaviraj A, Saha S (2015) Deposition of chromium in aquatic ecosystem from effluents of handloom textile industries in Ranaghat-Fulia region of West Bengal, India. J Adv Res 6(6):995–1002
125. Holkar CR, Jadhav AJ, Pinjari DV, Mahamuni NM, Pandit AB (2016) A critical review on textile wastewater treatments: possible approaches. J Environ Manag 182:351–366
126. Nakhate PH, Moradiya KK, Patil HG, Marathe KV, Yadav GD (2020) Case study on sustainability of textile wastewater treatment plant based on lifecycle assessment approach. J Clean Prod 245:1–15

HANDLOOMS: Unleashing Cultural Potentials

Marisa Gabriel

Abstract Since the beginning of human history, human beings have draped their bodies with different materials. These elements which soon became textiles play a fundamental role in communities and societies. Not only for its basic function of involving, covering and presenting the body but also as the expression of a community. Handloom textiles must represent the ones who wear but most of all, the ones who do these pieces. The fashion system in big cities is far from representing the values of the community or representing the values of individuals. However, the textile fabric, in particular the looms, weaves the stories of the women who do it; it weaves the identity of the community. The loom is a symbol of the community's social interlace. A loom is usually a place of calm and joy for women who practice this art; it is also a place of encounter, an excuse that joins and meets different women from the communities. And not just as a way of expression and liberation, but the loom also becomes the sustenance for their families. Women manage to earn their income, take care of their children, empower themselves and connect through the threads. The looms provide autonomy and sometimes enable a new way of organizing the families. That is why the loom becomes a central artifact in the courtyards of their homes. They are located in the center and are made with wooden posts, from trees obtained in the surroundings. In the same way, all the tools that this practice entails are made from natural materials of the environment; just as they learned from their ancestors. They carry out the whole process, from the shearing or obtaining of the raw material, to the curing, dyeing, spinning, and even the production of the finished piece. As mentioned before, this practice connects women with other women in the community, allowing them to establish ties, networks, and enhance their work. The importance of not being an isolated weaver but of assembling a network of weavers that can even allow the distribution of the product, the sale, and meet the quantities that may be required. However, how much of this technique survives from what was learned or taught by their ancestors? How do they balance the outside demand for colors, and materials, with the expression of their inner selves? Are in these pieces

M. Gabriel (✉)
Sustainable Textile Center, Paroissien 2680, 5th "B", C1429CXP Buenos Aires, Argentina
e-mail: mara@ctextilsustentable.org.ar
URL: http://www.ctextilsustentable.org.ar

M. Á Gardetti and S. S. Muthu (eds.), *Handloom Sustainability and Culture*,
Sustainable Textiles: Production, Processing, Manufacturing & Chemistry,
https://doi.org/10.1007/978-981-16-5967-6_3

traces of the ancestral character? How do they merge to sell the pieces and enter the national or international market? What are the challenges they face? Handlooms are a symbol and a product of the community's social interlace and it is the way to maintain a live spirit of the community and their beliefs. In many cases, handlooms today are a way of being, and also a way of living. At the same time, this practice might have been romanticized over the last years but is still missing a true sense of history, revalorization of this technique and the cultural heritage. The care for heritage implies preserving textile memory and disseminating its importance, as well as rescuing ancestral knowledge and practices of the weavers who still survive in this overwhelming globalized world, through the promotion and teaching of different handloom techniques.

Keywords Cultural heritage · Ancestral knowledge · Weaving · Textiles · Handloom · Women · Communities

1 Introduction

Textiles have the potential of being a vehicle and driver of culture. Furthermore, handlooms are full of meaning and could be the key in order to express, trespass, transcend, knowledge and cultural diversity. Inside communities, the textile represents an art, a language, the weaving of emotions, notions and concepts—the weaving of particular cosmovision of the world. Outside, it has the power of sharing a message, understanding and showing this particular worldview to others, because, the ones who do it, the weavers tend to put into the textiles their whole cosmogony; and at the same time, this is shared with the people who appreciate those textiles, and with the community. This chapter paper proposes and analyzes the handloom in Central Southern Andes of Argentina, its meaning through history and at present days. The reader will find in this chapter, a brief summary of the significance of apparel through ancestral communities and modern societies; the technique of handloom and weaving in the Andes, the meaning, the importance as a cultural and social practice. The chapter also includes types of handlooms, and the type of pieces made in this technique, and an analysis of the present potential and key challenges of the technique.

2 Roles of Apparel in Ancestral and Modern Societies

Clothing plays a fundamental role as a cultural sign; clothing has served and serves as a fundamental differentiating element of man within his culture. It can be defined and studied as a branch of archeology that is associated with artistic elements and accompanied by the approximations of different historical periods; as a cultural symbol that conveys meanings, and through the image of the individual, and the image of the textile or the garment, it becomes corporeal and evident. The dress is the biggest revealer of social functioning. It is the one that speaks the most because it is, at the same time, material, good investment and language [1].

The history of clothing is offered to us as an open book which allows discovering the most varied aspects of our historical past. It is difficult to determine when the human being decides to cover his body. However, from the most remote antiquity, the images captured in the various artistic manifestations and archeological reservoirs have shown that the dress was associated with the human being almost since its appearance on the planet. The act of dressing was accompanied by the search for differentiation, and implicit in it would be the desire for novelty and change that would give rise to that social phenomenon that we know as fashion.

By 1839, Balzac affirmed that for a woman the dress is a manifestation of the most intimate thoughts; it is her language, and it is a symbol. In this approach, clothing is presented not only as an explanation of society but it is also covered with a broader character, since it studies not only the social but must be studied in an intermediate dialogue between the body and the exterior, participating in the construction of the individual [2]. Alison [3] in her work, *The language of clothes*, mentioned that "if clothing is a language, it must have a vocabulary, a grammar like all the others languages. Of course, as with human speech, there is not a single language but many: some (like Dutch or German) closely related to each other, and others (like Basque) almost unique. And within every language of clothes there are many different dialects and accents, some almost unintelligible to members of the mainstream culture" [3] (p. 4).

Evidence shows that ancient civilizations not only dressed to protect their bodies and the need for shelter, for example, but also did it for moral reasons, and in an orna-mental way, to represent their tribe of belonging, their role within that community and decorate or beautify their bodies. There are remains that indicate and refer to the activity of textile art as early as the Paleolithic, 20,000 years before Christ. Even before the creation of textiles, the skin of the animal, their feathers and leather were used to cover the human bodies, and once textiles appeared as robes or cloaks, these materials continued being used by some of the people in the community or tribe. Therefore, clothes became a social practice. The accessories used are the needles, the buttons and the decorative elements that make clear testimony that clothing was an established social practice. Migeon [4] asserts that in the complex phenomena of artistic penetration, the textiles being portable in nature have been agents of transmis-sion of signs. Its influence on civilizations has been continuous by the communication capacity of its own language. It was an evident cultural sign, and the loom, the proof of it. At the end of the Stone Age, there are evidences of looms, built by horizontal tree branches, from which the fibers that make up the warp are held, and tensioned with stones at the other ends.

Furthermore, in this same line, Hughes [5] sustain that the fabric is present in our private life as well as it is an agent of collective, gestural, and symbolic expression, endowed with a communicational pregnancy. It is a mediator between the inside and the outside, between the intimate and the public. Textiles analyzed from its role as a vehicle and driver of the culture is full of meaning and could be the key in order to trespass, transcend knowledge and cultural diversity. Inside communities, the textile represents an art, a language that weaves emotions, notions and concepts. The ones who do it, the weavers, tend to put into the textiles their whole cosmogony; and at the same time, this is shared with the people who appreciate those textiles, and with the community.

As mentioned by Margarita, a woman with indigenous descendants from Cata-marca: "Symbology in handloom pieces are an icon underdeveloped as something of our own, that is sometimes hidden for us, but not so, for the one who finds it, and in handlooms you may found it, and she ends, no-one is a prophet in their own land" (Margarita, interviewed January 2021). If the textile is a piece full of significance, the handloom might be the instrument, the place and the object that allows magic. Handlooms weave the stories of the women who do it; it weaves the identity of the community as we will see in the different interviews made for this paper. But, how can we describe the handlooms used and still use in the Andes? Which were the pieces made in those looms? Which was the technique? How does this textile speak for the weaver? How does it express the deepest thoughts of their souls?

3   The Art of Weaving

As mentioned above, textiles are, and reflect, an important part of the evolutionary process of humanity from early times to the present day. First, textiles appeared as protection against the environment, and then they were soon transformed into an artistic manifestation, into ideas to mark ethnic differences and belonging, into elements of high value within religious, military contexts, etc. Through the centuries, the clothings evolved according to the influence and requirements of each era; those could be moral, religious, social, economic and even political orders. However, hand-looms did not suffer this transformation, at least not in the way of doing; those changes might be reflected on the uses, the offer, and demand of the textiles, but very little in the making process.

Handlooms and traditional textiles speak of culture. Artisans show up their legends and idiosyncrasies, their geographical and ecological environment, their relationship with nature of which they feel an active part. The weavers embodied in the looms their life, which is also the ancestral memory of their peoples, their philosophy and identity; they tend to preserve their customs; as well as their own symbolic conception of the world and the universe from birth to the moment of death. Hence, we can rescue the exotic art expressed in archeological, ethnographic and historical fabrics: either as tapestries, traditional fabrics or as elements of clothing used in different daily activities, social, religious, etc., and also, current textiles. Therefore, it can be concluded that textiles are one of the elements participating in the evolutionary process of humanity.

According to Corcuera [6] (p. 29), "Fifteen thousand years ago, the early hunters found in South America a vast environment. The high complexity attained by weaving in the Andean cultures is the result of a process dating back to thousands of years". In this sense, handloom textiles are a faithful reflection of this vast environment, both a representation and a product of culture. The weavers in the South Central Andes suggest that this wider process involved more than the technological developments in looms and instruments, or in weaving structures and techniques. It was rather the historical integration of these developments into social sequences and the cascade effect of these wider social activity streams that articulated the elements of these broader Andean territories into an evolving whole.

In textiles and handlooms, in the art of weaving and knitting, appeared the essence of the spiritual life of families, and beliefs were also part of the interlace. The environment not only gave them the materials but also in the environment they find inspiration and simulation. As an example, there are the testimonies of women who as a child, learnt from their mothers and grandmothers, to make a yarn with abandoned spider looms, placing it on their wrist so that the virtue of the little bug, it was transferred to the woman. They were asked to take a good look at how the spider works with its tiny hands and so delicate that they could do it the same way.

"These fabrics refer us to a more essential identification: they remind us not only of a more natural lifestyle, but also of nature as an essential condition for all life to develop. This amazing mountain from which these pieces come to us is a source of union and contact with our deepest roots, not only those of our history, but also those that connect us with our natural origin, our origin as inhabitants of an uncontaminated planet" [7] (p. 25).

Handlooms textiles symbolize the communities who practice their identity and memory of the origin: grandmothers used to do that. Handlooms were their life, their craft, their job; just like the chair, the table, or the bow was for their grandfather. In most traditional homes, you find that still today, women knit in handloom a piece for each son or daughter when they get married or when they leave home. This was, and still is, for them to take home within, the memory of their home, their family, and their mountains. It is not only to shelter the body but to shelter the soul with the memory of their home and their land. The handloom pieces were embroidered with nature motifs (as seen in Fig. 1).

"What I never want to be lost is the transmission of the inheritance that mothers left to their children: the fabrics, ponchos, blankets that mothers left to their children. Those pieces, for example, the *poncho*,[1] spoke: I protect you from cold and heat with my whole being; and I will accompany you wherever you go" (Margarita, interviewed January 2021).

---

[1] Poncho is the Mapuche word for garment. A typical traditional cloth used in the Andes.

Fig. 1 Handloom piece
embroidered from
Tinku-Kamayu Cooperative[2]

4   The Handloom in Argentinian Andes

4.1   Before and After the Appearance of Handloom

Before the handloom appeared in the Andes region, the technique used to make
textiles was the Sprang. Sprang is an ancient method of textile construction that is
very similar in appearance to weaving, but it has a natural elasticity and it is entirely
warp-built. Archeology shows that the sprang predates handloom and weft weaving.
It is similar or similar in appearance, but technically very different since it is a
single thread, and the loom always consists of two threads. The sprang technique is
universally considered and one of the first used by humans. It seems it was present in
Europe since the Iron Age, although it was only identified in the nineteenth century
[8]. Its technique is very simple, as it can be done with a frame and some support
rods. Unlike most techniques, the sprang starts from the center and grows to the

---

[2] Tinku-Kamayu Cooperative is formed by a group of organized women gathered to work. They are
expert spinners and weavers who transform llama fiber, vicuña fiber and sheep's wool into yarns
and then design garments that preserve the naturalness, softness and colors that characterize them.
They make clothing with a typical and artistic touch of the environment. Their emotions of humility,
patience and creativity appear in each piece and are the essence.

sides. The use of the rod or stick in the center is what prevents it from unraveling, and by crossing the thread the final design is achieved. Needles with or without an eye were the most widely used instrument to make this textile.

The use of a loom, however primitive, means a significant advance in the history of weaving because by utilizing a heddle rod, less movements are involved. According to Carballo [7] based on the data available to date, regarding the appearance of the handloom, an imprint of this instrument was found in the Valdivia culture, Ecuador, between 3000 and 2500 BC. Those who still live among the mountains and keep practicing handlooming know that the materials that were traditionally used were the most commons of daily life, not only to build the handloom and the necessary instruments but also to weave. Those materials allowed them to pour their aesthetic feeling into fibers they obtain from the surrounding environment. They spun everything they found in the forest that could be woven. From these encounters between people, tradition, and the environment some of our most refined popular arts were born in terms of material and realization. Regarding the handloom, Dellepiane Cálcena [9] points out a simple instrument, a frame of four tightly tied poles (see Fig. 2). A simple instrument from which complex techniques are obtained is one of the characteristics of pre-Hispanic textile cultures, which is why they are also an example of complex handmade "industry". These looms allow a double-sided fabric to be achieved with the support of floating warp threads that, sometimes, usually skip up to three wefts for the construction of the design.

Traditional handloom weaving is still practiced in Argentina from the Puna, Chaco and the extreme northeast to Patagonia. It lasts in the Creole population and also among almost all indigenous groups. The fibers used are wool, vicuña, guanaco and llama. Sheep wool was introduced by Europeans but today is the fiber that they use

Fig. 2 Vertical handloom.
Source Prepared by the author

Fig. 3 Handloom in vicuña
wool

the most, and also cotton, another crop promoted in those years by the colonizers. In
the beginning, the quality of cotton in Catamarca was striking, but then the industry
in the country lagged far behind the Egyptian, the Indian cotton or the United States.
At that time, among other fibers that were used and that are less and less used were
the wild coyoyo silk from butterfly cocoons, palo borracho cotton, palm fiber, brava
nettle, goat wool, and even dog hair. Nowadays, the most common threads applied
to handlooms are wool and cotton, or industrial mixes, this accelerates the process
but has its disadvantages or potential risks. See in Fig. 3 a piece knitted in vicuña
wool, one of the most exquisite fibers of the Andes.

It is important to mention that handloom requires the artisan, the human being,
and as a product of this person, the mood for example will be transferred to the loom.
The fiber might be an industrial thread, but the piece will always represent the artisan
who made it. The mood of the artisan is very important. If the weaver is not in the
mood, everything can be reflected in a tapestry on a cloth, and it cannot be otherwise;
it is the artisan and its history, it is the artisan and its life that surrounds him, the
artisan is not a machine, and cannot work like a machine (Margarita, interviewed
January 2021).

4.2   Handloom Frames

Handloom is characterized for having two systems of threads, warp and weft; for this,
two types of devices are used. The first, a fixed or adjustable quadrangular frame,

sometimes with legs, which may or may not be transportable, can be improvised with four stakes and two crossbars or with the four legs of a table or an inverted cot. In it, the warp will be assembled, and the pieces are manufactured.

The looms have one or more healds to separate the threads from the warp. In Argentina, several models are used, most of them indigenous. All indigenous looms lack pedals and threads must be moved by hand; there are vertical and horizontal models. These artisanal looms are classified into two main types: vertical and horizontal. The former are wooden rectangles held vertically on a base; sometimes they have a table as a seat, added to their vertical beams. Among the Mapuches of Argentina, the presence of a vertical loom with a horizontal warp has been documented; it is also known that they used a horizontal loom supported by four stakes. The latter is made up of wooden frames that contain needles or meshes through which the threads to weave pass through, mainly cotton, camelid or sheep wool. Likewise, the so-called backstrap loom is still used, a very primitive but easily transportable system.

There are three forms of vertical looms that will be detailed: two consist of two posts or pitchforks, on which two crossbars are fixed that support the warp. In one, the shaft of the heddle is supported on two posts, known as horcones, on which two crossbars that support the warp are fixed. In one, the shaft of the heddle is supported on two additional posts; in the other, the heddle hangs unsupported and may lack a stem. These vertical looms must surely have originated in Andean cultures, from which they spread to the Mapuches (Araucanians), Tehuelche and the Caqueños. The third vertical loom consists only of two vertical posts between which the warp extends. It is used to weave girdles by the Mapuche and Tehuelche, and it shows a certain Amazonian influence. The Araucanian looms used by the Creoles are usually frames made by carpenters.

4.2.1   Type of Handlooms

Vertical handlooms

- The vertical frame loom: two thick poles of almost 2 m that end in forks are driven into the ground as stakes, and the distance between them must always be greater than the width of the cloth. Two wrappers are tied to them in which the fabric is woven (as shown in Fig. 2).
- The backstrap loom, sticks looms, girdle loom: whoever weaves uses his body as an instrument. After passing the warp threads to the thumb, along and on the extended leg, which marked the extension of the fabric, and turned it into a somewhat personalized element.
- The belt loom or waist loom: it was used to weave blankets, for example, long fabrics, which are tied to a tree or wall, and the other end was girded at the waist by means of a belt made in rural areas called, *tiento*. It was allowing the expansion of the technique adapting two parallel poles to tension the warp threads. And so, wooden boards began to be used at the extremes of the canvas (see Fig. 3).

Fig. 4 Waist loom, or belt
loom. *Source* Prepared by
the author

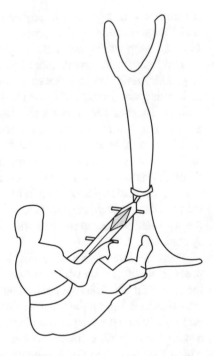

The vertical frame loom, although it is similar to the previous one, the poles that act as stakes are not placed fixed but tied with fibers to the beams of a branch or hearth or only approach a wall. These poles that measure between 1 m and 1.80 m are placed at a distance of the measure of the piece to be woven, and the weaver sits on the ground in front of the loom or on a short-legged bench, which is rather used to keep your torso upright (Fig. 4).

Horizontal handlooms

– Four-pole horizontal loom: used to weave all kinds of fabrics, this loom is made by planting four long rods perpendicular to the ground (two rear and two front) to hold the wrappers so that the fabric to be made will be parallel to the floor and very close to it. When the arms need to be stretched too far, the weaver rolls up the fabric and brings the back stakes closer. This instrument is currently used in Argentina and the Bolivian Puna. Small loom to weave fur, overclothes or sweat suits constitutes a variant of the frame loom, in which four fixed stakes hold the envelopes, thus forming a frame of the size of the desired fur. At an ethnographic level, in Argentina it is used in the provinces of Catamarca, La Rioja, Tucuman, San Luis and Mendoza (see Fig. 5).

In every handloom there are instruments or tools used to strengthen the fabric and differentiate them, the shovel, of indigenous origin, and the comb that arrives with the Europeans. From this, it is that scarfs, shovel ponchos or comb ponchos are made (see Fig. 6, a scarf made by Rosita, one of the weavers from the Andes that

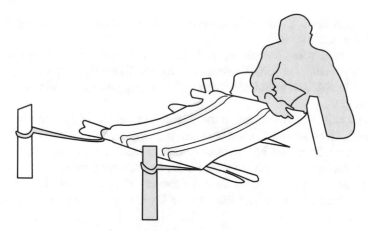

Fig. 5 Horizontal handloom. *Source* Prepared by the author

Fig. 6 Handloom scarf made by Rosita

was interviewed). The blade is a piece of wood that is usually 70 cm long, its height is 10 cm and it has a bevel that goes over one of its lengths. It is made of heavy wood and its purpose is to tighten the thread that it is weaving with each pass. It is also the blade that before tightening the weft contributes to open the passage of light or shed. The weaver usually has several blades of different sizes to achieve different results. The comb instead, brought by the Europeans, is a basic element within the looms and allows greater speed, so in most cases the blade was replaced.

- Mathina or comb, a tooth-shaped bone spatula, used to adjust the weft in a similar way to that of a comb.
- Ruqqui, ruki, rokey or punzon, made with a hollow metapodium of guanaco or another camelid, is a pointed instrument that recalls the shape of a dagger, and the weaver wields it in a similar way to pass it through the entire warp and separate its threads.
- Callhua kallwa or shovel, when the fabrics had designs in their warps, to adjust the weft, Quechua and Aymara used and use wooden or bone shovels of different dimensions. They were known as kallwa. They are flat, elongated instruments that have the length of the width of the tissue. Its ends are pointed to a point, which facilitates its introduction between the threads. Generally, woods such as palm trees or those of greater hardness like carob were sought for their manufacture.

4.2.2   Handloom Designs and Garments Made in Andean Handloom

The looms may vary according to their material, design, and also their color. The dyeing techniques that occur in the spinning prior to weaving used to be with natural dyes; nowadays, the color is usually achieved from anilines. The colors are interspersed; one of the most used decorations in the handloom is usually called a comb, which consists of longitudinal lines formed by a series of transverse bands of two alternating colors. To obtain it, it is necessary to weave half a turn of one color and a half of another, which is easily achieved by weaving with two balls of wool of different colors and interlacing the threads on each crossbar. At the end of the warp, a layer of threads of one color and another of another remain at the level of each list. At each crossing of threads, the opposite color passes to the front face, and thus the alternating bands of colors are obtained (see Fig. 7, comb technique).

A tubular double-sided technique can be used, which is widely used in the manufacture of girdles with the vertical loom with two stakes and the vertical loom with a

Fig. 7 Comb color technique in various colors from a piece of Tinku-Kumayu

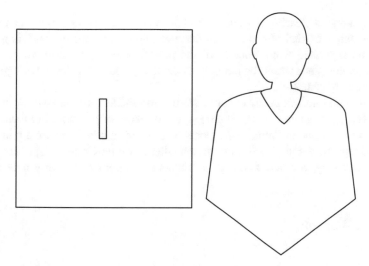

Fig. 8 Poncho representation. *Source* Prepared by the author

vertical warp. Decorative bands for ponchos are also achieved through this technique. However, the technique widely used in ponchos is the warp floats double-sided. It is characterized by the fact that some warp threads remain free in non-corresponding places on one or the other side of the piece, sometimes jumping two or more passages of the weft. In this technique, weavers work with three or four planes of threads. Two planes always intervene in the crosses and the rest will form the floating decorations on one or both sides of the piece (see Fig. 8, a basic poncho representation).

Among the characteristic pieces that come from the manual loom is the poncho, very widespread in the countries of America, with a great presence on both sides of the Andean massif, and which undoubtedly makes up the cultural heritage of the region ([8]: 29). The poncho is a pre-Hispanic antecedent of these lands, a wide rectangular garment loose at the sides and with an opening in the middle [8]. In 1969, researcher Huepenbecker A.L., following King's review, analyzed 280 poncho-style pieces in the Andes and found that this garment was not only a coat but also served other functions. Its apparent modesty was linked to a complex belief system. The relationship between Andean cultures and those of Central America is evident in the different ponchos. For example, Chichén Itza, a theocracy that based its power on astronomical knowledge [10], were the intermediaries between heaven and earth, were privileged within society, and the blue feathers of the ceremonial ponchos had a strong symbolic meaning among them. In the Andean world, the vicuña is the strongest representation of the solar myth. The father Inca, the one who gathered all the power, like this camelid, depended only on the Sun. The poncho was the garment worn by the Indians, and the vicuña, in particular, could only be worn by the highest rank.

The poncho is clearly associated with equestrian life and the Argentine pampas. The researches carried out in the last decades allow us to affirm that this garment cannot be separated from features of the pre-Hispanic past, although its greater diffusion takes place when the indigenous person becomes a rider in the eighteenth and nineteenth centuries.

At the same time, not only ponchos were made and are still made today in hand-looms. Each village characterizes some type of textile or garment. For example, in Catamarca, you find in Santa Maria ponchos or tunics made by a woman; in Belen, you find tapestries and ponchos and in Andalgala you may find bedspreads. At the same time, today, you may find bags, backpacks, cushions and other products.

5 The Weavers

A good blanket is like a poetry built on rhythm and time, or, in other words, cadence and time incite the poet to compose under his real aspect

(Corcuera [8]: 83)

In Argentina, the weavers of Belen, Catamarca and Santiago, Santiago del Estero, are well known; however, the handloom extends throughout the country in most rural populations and abroad the limits of Argentina in the entire Andes. The *telera* is the name given to weavers across the Andes in Argentina. In the words of Margarita, "I learned from my mother and my grandmother, helping them to make the threads, washing yarns, joining fabrics. At the age of 14 we got together to weave and spin the yarn. I liked going to visit my grandmother because I liked to see her, her sisters and daughters knit, laugh, talk, I loved being with the family, sharing the threads. Weaving always united the family. Women in the weaving, and men in the tools, and fibers needed. This technique is in my heart, my body and in my hands" (Margarita, interviewed January 2021).

The importance of the domestic textile industry was, and it is today, reflected in the structure of society in rural areas. The handloom was used to attract women between the ages of 16 and 29, not only to continue its lineage but also to achieve economic independence, generate income and help their households. At first, the control of weaving by women configured a situation of unique characteristics. While it meant that they were enabled to participate in employer–client relationships, the weavers needed according to Hermitte and Klein [9] "support and mediation from employers who occupied a privileged status. We then find a class system with a minority that owns farmland and engages in commercial activities and a large majority of workers dependent on them. The independent weaver was almost non-existent if we excluded the ladies of the high position who were in charge of this task in addition to having the labor of laborers" ([9]: 35). In the words of Rosita, handloom weaver in Catamarca, "Knitting day and night to breed, knitting was my company, my livelihood, my psychologist because I spoke alone while I knitted. Taking out the fabric was a triumph, an achievement, and it gave me joy, a process was closed, with each fabric the soul was healed a little, and it was the beginning of a new one, the next one" (Personal conversation, January 2021).

The exquisite technique of handlooms is precise, and only years can give the expertise to the weavers. "The *telera* cannot be wrong, its hours are not counted; She will be aware that the warp is correct, the plot fits well, and her mind will be disciplined for the mathematical calculation, which implies achieving the design that you want to achieve" ([8]: 81). And she adds, "Colors and shapes, memories and innovations are part of the current expression of these textiles, which come to us like a light wind to the city" ([8]: 81). Virtuosity and expression are always present in the sensible, exquisite pieces of handloom. Historically, with the arrival of the trains, the *teleras* absorbed all the novelties of the turn of the century, and under the influence of Victorian designs, their works were filled with curves and flowers. Over the years, the strength of those first and original designs of the earth was diluted, and the art of *teleras* practically disappeared into oblivion along with the mountains that inspired them for at least a long millennium [7].

The process of creation for Margarita was described as an inspiration based on nature. She mentions: "Before creating, I go up to the mountain, I go to meet nature to inspire me, I let myself be carried by the air by the wind, the water, the earth and the plants, and the respect of my ancestors, because at that moment of remember them in their daily tasks in their lives so simple so great that they left a legacy written even on the stones that are washed with the dawn. It is so mystical so beautiful that I feel it, and if I focus on creation to create what I believe is not mine, it has an owner, and that gives me freedom" (Margarita, interviewed January 2021). In this same line, Rosita added, "The handloom is my life, it is my soul, without fabric without threads I am nothing, my house is full of threads and colors, although I do not live from weaving because I am a teacher, weaving is the most important thing in my life" (Rosita, interviewed January 2021) (Fig. 9).

As mentioned by Corcuera [8], the girls are heirs to the ancestral art of the perfection of spinning, the art that amazed the Spanish visitors, at the end of the eighteenth century, when they came to a town of weavers in Santiago. Over time, those encounters were bearing fruit, since it was not only sought that those memories were not diluted, but also that the weavers gain and had autonomy. It is the difficult transition of many of these girls to modernity. They became aware of the pre-Columbian past with great respect. At this stage, the aim was to underline the autonomy of these small towns. For the first time, the weavers were faced with a challenge: to reach the market of the cities without losing their identity. These young women faced with internal tension, which causes the decision between roots and uprooting, the greatest contradiction that our modernity generates among these communities, and this internal struggle between roots and uprooting, rural areas and big cities, might be one of the important issues found in the communities around handlooms still nowadays.

Fig. 9 Open poncho
embroidered Tinku-Kamayu

6   The Present of Handloom in Argentinian Andes

As seen before, rural weavers are rooted in an environment in which contact with nature is extremely strong, not only because of the adversities of daily life but also because of the gift that nature gives them: the possibility of living in a place where dialogues still exist. The noise of the loom, the wind, the birds, and the participation of daily life that they return to the loom. Thus, this cosmovision is represented in these pieces that come out of the handlooms. At the same time, in handlooms, they find calm and joy, while spinning the weavers laugh and share. But what happens today with these iconographies? With this heritage, do the weavers continue to represent these issues in their pieces?

For Margarita, "The fabric is an encounter with oneself and it is unique when it is shared with another, life is shared from thread to thread, and the fabrics are made faster. The experience, for example, at the Center Aurora NOA,[3] was a blessing, knitting with other classmates, and every knitting encounter was a party" (Margarita, interviewed January 2021).

---

[3] Aurora School, Aurora de un Nuevo Mundo, is a Technical Formation Centre in Crafts that since 2005 provides with formal education to the community of Santa Maria, Catamarca, by offering a Technicature in three specialties: Metalwork, Ceramics and Textile. Our main objective is to promote the revaluation and rescue of the original cultures through the pedagogy of reciprocity, promoting social mobility and improving the employability of the community. (https://www.centro auroranoa.com).

A craft of such aesthetic value and as rich in forms as our traditional weaving, which fulfills such necessary functions as shelter, transport, and clothing, and which is usually an appreciable source of income in the meager family economy of certain rural areas, faces now a number of adverse factors. To begin with, it is an under-valued product; the traces of the cultural diversity and the symbols that these artisanal pieces merge have been lost over the years and the modern societies. For example, today, in rural areas and villages, it is common to see some of these historical traditional pieces found as closures of chicken coops or covering animals in the barns. At the same time, to mention the fibers and the materials used, "Today fabrics are made and cut to produce other things is no longer the same. Cutting a fabric in the past would have been an aberration" (Margarita, interviewed January 2021). This is a product of a system of fashion that standardizes all, and there is no place for traditions, and to hear different voices, there is no place for different arts and perspectives. Therefore, if doing a handloom piece take months, and then ends up in a chicken coop, or being cut as a standardized industrial textile, a young woman will try to find different opportunities and jobs because the one they saw their mother and grandmothers did, they see how demanding and laborious it is, and there is no possibility of valuing it, and selling it as an art piece, or cultural piece.

"The children of weavers no longer want to leave that sacrificed life they saw their mothers and grandmothers lived, they have different dreams, and in handlooms they find a poorly paid job, where women even leave their bodies in the handloom pieces and that is not recognized and is not well paid" (Rosita, interviewed January 2021). And she added, another big issue around handlooms are breathing problems, all the fluff chinstrap that goes to the body. Rosita remembered blowing the nose and seeing tinctures and fluff. It is a difficult job, not very healthy and not well paid, so the youngers do not want to continue it.

At present, and in the last 20 years, young women in rural areas have to decide between traditional art and a way of living, and a modern way of living. In this decision, they have to balance the historic understanding of their own ancestors and families and modern societies. And it is very hard to value this technique and ancestral wisdom when there is any financial support and handlooms are not valued as a cultural-historical piece. It is common to see ponchos, sweaters, bags, cushions and different products, where the handloom piece, instead of being knitted exactly, without any residual material, is being knitted as a big rectangle, from where then, each pattern is cut-off. This, not only changes the basis of handlooming but also gets near an industrialized process that could be imitated (see Figs. 10 and 11). However, a woman in rural areas do this, because their work and art is undervalued, and this way open new possibilities and accelerates production. In Fig. 12, it can be seen, a different way of constructing an artisanal handloom coat, finishing it with details in crochet, another handknitting technique used in Latin America; and in Fig. 13, another embroidered handloom poncho, with flower details.

Somehow, in order to solve some of the present issues of these techniques being lost, industrialized fashion, designers and entrepreneurs are trying to make a fusion between traditional handlooms and modern garments. In this way, there are different possibilities of approaching weavers and communities to try to preserve these tech- niques while opening new possibilities and new markets for weavers. "In our South

Fig. 10 Handloom artisanal bag

Fig. 11 Handloom cushions

American countries, we find designers who fuse original and traditional elements with their own designs and contemporary language in their creations, thus highlighting the originality of our native culture. In these designs they rescue myths or legends by going to the visualization of oral traditions" ([7], p. 186).

One of the main issues around this topic is that handlooms tend to be sold in an anonymous way, without telling the name of the artisan behind the piece, and this unfolds different issues like, for example, the weavers tend to stay apart from the selling or distribution of their piece, and there is no real connection with the people who buy each piece. At the same time, the one who buys it loses part of the history and culture of the piece, being aware only of the part of the story the entrepreneur and the brand want to be known. Sometimes this has to do with egos; brands don't want others to know exactly where and how artisanal products were produced, not to be copied or skipped in the offer demand equation. However, the most important

Fig. 12 Detail of handloom coat, with a mix crochet technique

Fig. 13 Embroidered handloom poncho

theme is to try to embrace sustainable conception. Valorizing the artisanal behind each piece, and building a strong relationship with the weavers should be the first step for a brand trying to sell and preserve artisanal crafts. Handlooms depend on the artisans and the people that give life to handlooms pieces.

In the words of Margarita, "Handloom is about living, motivating ourselves, valuing ourselves, believing that we can go beyond dreams, and it is very difficult when the pieces are not valued as what they mean for each of us. We are people of the community, who say little and expect little, as is expected the day and the other day, humility is a very strong characteristic in this area. And after years of suffering forgotten in the mountains, hope, or dreams, are no longer so great, I only dream that this, the essence of handloom, is never to be lost" (Margarita, interviewed January 2021). And she adds, "There are always people who want to learn, but as an art, not to transcend. Weaving is art for the soul, and art that serene the heart, it is a therapy to find oneself" (Margarita, interviewed January 2021). New generations have the possibility to exploit this resource, that is native; it is ancestral, and it is full of meaning; they can really make a difference, or if we continue undervaluing it, it may disappear.

7 Conclusions

As mentioned before, handlooms are a symbol and a product of the community's social interlace and it is the way to maintain a live spirit of the community and their beliefs. After analyzing different perspectives and interviewing different artisans, the conclusion that unfolds is that in many cases handlooms today are a way of being, and also a way for living. Selling those pieces is for the artisans a way to stay alive and at the same time, as any job could be, is a way to organize and sustain the household. In the particular case of handlooms, the possibility to live in the rural areas and develop their crafts without having to leave from rural areas to villages or cities near is very important. At the same time, this practice might have been romanticized over the last years but is still missing a true sense of history, revalorization of this technique and the cultural heritage. It is very little what is known about the history and symbolism of handlooms. Brands nowadays, which work with handlooms, tend not to be compromised with the community and tell or show the part of the story they want to tell, is common to see many artisanal crafts being sold from the Andes, sometimes they mention the province where it was done, but rarely the artisan. And there is being lost the potential to hear those artisanal voices that could accelerate sustainability.

In order to transmit, conserve and preserve this wisdom, the selling of the products should be revalorized, as the "product" being recognized as a cultural product. At the same time, studying deep on these themes and creating documents would be part of the preservation, and a necessary step to put this knowledge in service for teaching everything around how cultures are reflected on artisanal craftsmanships; in this case, handloom textiles. The care for heritage implies preserving textile memory and disseminating its importance, as well as rescuing ancestral knowledge and practices of the weavers who still survive in this overwhelming globalized world, through the promotion and teaching of different techniques and original methodologies.

References

1. Coquery N (1998) L´Hótel aristocratique. Publicaciones de la Sorbona, Le Marché tu luxe a Paris au XVIII siecle. París
2. Honoré B (1964) Una hija de Eva Editorial. Nauta, Madrid
3. Lurie A (1992) The language of clothes. Twayne Publishes, New York
4. Migeon G (1929) Les Arts du Tissu. H. Laurens, Paris
5. Hugues P (1996) Tissu et travail de civilization, Rouen, Eduitions Medianes, Paris
6. Corcuera R (1991) Herencia Textil Andina, Andean Textile Heritage. Fundación CEPPA, Buenos Aires
7. Carballo B, Paz R, Boschi L (2005) Teleras, Memorias del monte Quichua. Ediciones Arte Étnico Argentino
8. Corcuera R (2017) Ponchos de America, de los Andes a las Pampas. Fundacíón CEPPA, Buenos Aires
9. Dellepiane Cálcena C (1960) Consideraciones sobre la tejeduría de una comunidad de origen araucano: Azul, provincia de Buenos Aires
10. Alcina Franch J (1965) Manual de arqueología americana. Editorial Aguilar, Madrid

The Influence of Culture on the Sustainable Entrepreneur: An Investigation into Fashion Entrepreneurs in Saudi Arabia

Rana Alblowi, Claudia E. Henninger, Rachel Parker-Strak, and Marta Blazquez

Abstract Research on sustainable luxury fashion has increased markedly in the last couple of decades, allowing academics to combine entrepreneurship and sustainable development. Sustainable luxury is a focus on cultures, values, traditions and needs. Additionally, socio-cultural norms encourage female fashion entrepreneurs in Saudi Arabia to be more sustainable by employing social and cultural norms in their designs. Fashion plays a key part in Saudi culture. There are numerous garments that have remained popular for long periods, as a result of the exclusive fabrics being used in their creation, the cultural heritage of the country and the artisan handicrafts employed. Traditional handloom garments are acknowledged as an important part of the sector. In fact, one of the goals of Saudi Vision 2030 which encourages sustainability states that it should "establish itself as a dynamic leader in fashion by preserving traditional national customs while developing the next generation of Saudi designers, and sustainable practices that can compete on a global level". The aim of this particular chapter is to explore the part that female entrepreneurs play in the field of luxury fashion industries and in striving to achieve Saudi Vision 2030 in relation to sustainability and reviving fashion heritage. This is an exploratory chapter. The authors have used academic literature together with qualitative (semi-structured interviews) interviews conducted with Saudi female entrepreneurs with the aim of attaining a greater understanding of luxury and opinions concerning cultural heritage and sustainability.

R. Alblowi (✉) · C. E. Henninger · R. Parker-Strak · M. Blazquez
Department of Materials, University of Manchester, Oxford Road, Manchester M13 9PL, UK
e-mail: r.alblowi@psau.edu.sa; rana.alblowi@postgrad.manchester.ac.uk

C. E. Henninger
e-mail: claudia.henninger@manchester.ac.uk

R. Parker-Strak
e-mail: rachel.parker-strak@manchester.ac.uk

M. Blazquez
e-mail: marta.blazquezcano@manchester.ac.uk

Department of Materials, University of Manchester, Manchester, UK

Prince Sattam Bin Abdulaziz University, Al-Kharj, Saudi Arabia

© The Author(s), under exclusive license to Springer Nature Singapore Pte Ltd. 2021
M. Á Gardetti and S. S. Muthu (eds.), *Handloom Sustainability and Culture*,
Sustainable Textiles: Production, Processing, Manufacturing & Chemistry,
https://doi.org/10.1007/978-981-16-5967-6_4

Keywords Sustainability · Luxury fashion · Female · Saudi Vision 2030 · Cultural heritage · Indigenous crafts · Entrepreneurship

1 Introduction

The Saudi government recognises the need for a different direction regarding domestic growth in the shape of an economy that is more knowledge-based. It should be mentioned that innovation is one of the principal driving forces behind a knowledge-based economy [1]. In the last decade, Saudi Arabia has actively tried to move away from its oil dependency and towards establishing an economic business model that is more reliant on non-oil products, which is reflected in their Saudi Vision 2030 [2, 3]. As a result of Saudi Vision 2030, the country has seen a raft of initiatives pertaining to design and fashion, all of which concentrate on a way of life that is more sustainable and ethical, in addition to a developing youth culture that is propelling an aesthetic that has been emphasised as individual, extraordinary and exhilarating. Saudi Arabia has a highly respected and productive fashion industry within the Gulf region, and with plans on being less reliant on oil-based products, it may not be surprising that plans are being developed to expand with the intention of improving its economic prosperity [4, 5].

One of the goals of the Saudi Vision 2030 is *achieving environmental sustainability* "by preserving our environment and natural resources, we fulfil our Islamic, human and moral duties. Preservation is also our responsibility to future generations and essential to the quality of our daily lives" [3]. Saudi Arabia has a rich history of clothing and textiles, despite the fact that it is a relatively young country [6]. Traditional dress has a vibrant history and plays a significant role in the lives of the Saudi community [6]. For instance, each region in the country used to have its own unique style of clothing and textiles that were handmade by way of weaving, embroidery, sewing and dyeing. These distinct styles conveyed belonging to a geographic and historic location as well as an individual's standing in society [6–8]. The clothing that is traditional to a particular region represents a key characteristic of that particular area's cultural identity [8], and also helps to create a visual connection to the country's rich cultural heritage more generally, which is gradually transforming and shaking off the shackles of the past [8].

Regarding Saudi Vision 2030, one objective relates to protecting the country's national identity by way of preserving its cultural heritage in traditional costumes [9]. Although Saudi Arabia may have appeared careless in applying sustainable development in the past [10], with the introduction of the Saudi Vision 2030, sustainability becomes a focal point in that the government has thoroughly considered sustainability and how it must be encouraged and incorporated into the (fashion) industry, by way of promoting new projects surrounding cultural heritage and sustainability. The vision associated with the Saudi fashion industry is to become "a flourishing of arts and culture across Saudi Arabia that enriches lives, celebrates national identity and builds understanding between people" [11]. Although fashion within Saudia

Arabia has changed and become increasingly more global, in that designs are often inspired by other cultures, places, colours and other ideas of aesthetics. Some aspects remain the same, in that it (fashion) plays a crucial role in Saudi society, as it reflects the diversity and variety of Saudi society, its regions and history [4]. The increased openness of Saudi society is also reflected in the fact that Saudi Arabia is a key market for luxury products including fashion and accessories and, moreover, that it is the second-largest sector with respect to luxury commodities in the Middle East after Dubai [10].

In line with the Saudi Vision 2030, fashion organisations in Saudi Arabia are increasingly aware of sustainable measures that need to be implemented, and become more aware of the role sustainability plays in other markets (e.g. [12–14]). In 2019, Fashion Futures took place in Riyadh's Cultural Palace to "help Saudi Arabia establish itself as a dynamic leader in fashion by preserving traditional national customs while developing next-generation Saudi designers, manufacturers, education providers and sustainable practices that can compete on a global level" [15]. Thus, it is essential to appreciate how businesses and governments could collaborate in order to reduce consumption, thereby encouraging customers to make choices that are better for the environment (natural and social) [13].

It is worth stating that several new luxury entrepreneurs (globally and in Saudi Arabia) are developing remarkable perspectives on sustainable development and, furthermore, that both emerging luxury entrepreneurs and current luxury brands are working incredibly hard to incorporate sustainable development into their brands and collections [16]. This chapter contributes to Saudi Arabia's ambitious vision for the future. Likewise, the research will identify the changes that have occurred in regard to sustainability to help the implementation of the Saudi Vision 2030. This new trend across the country is helping to increase awareness of sustainable fashion within Saudi society with respect to every part of life, the importance of the profession as well as the craft and cultural heritage, for the reason that achieving the Saudi Vision 2030 can only be accomplished by educating people and fully preparing them for the labour market. Thus, this chapter poses the following questions:

Q1: How do Saudi female entrepreneurs understand sustainable luxury?

Q2: How can they foster Saudi Vision 2030 to combine sustainability and heritage?

2 Sustainability and Luxury Fashion

Luxury is dependent on cultural, regional or economic circumstances that convert an everyday item into something that may be seen as an indulgence [17]. Sustainable luxury cannot just be seen as a conduit for greater reverence with regards to social development in addition to the environment, but likewise will be a synonym for art and culture besides the creations of various peoples, preserving the legacy of local expertise and ancestral cultural heritage [18]. Luxury creates value as a result of objective rarity, which could be due to the use of rare/precious materials (e.g. gold, platinum) or to unique craftsmanship (e.g. design). Protecting outstanding crafts-manship enables luxury's rarity to be created. Moreover, luxury is often associated with quality, due to the price tag attached. As such, luxury items are often perceived to be more durable, which in turn can help reduce waste, resulting in a positive effect on environmental sustainability [17]. Luxury products can have a direct positive impact on the production and consumption processes, for example, by creating job opportunities (e.g. craftsmanship), as well as through taxation. At the same time, it can also have spill-over effects in terms of creativity and innovations. The latter can be evidenced through, for example, more mainstream brands "copying" luxury designs.

With the aim of monitoring this increasing research interest regarding sustain-able luxury, Athwal et al. [19] issued the first literature review to take into account sustainable luxury marketing. By way of arranging the existing literature within this field in conjunction with international and cross-cultural concerns, their work provided important insights pertaining to theory and practice. It is worth noting that the need for luxury maintains the maintenance of these skills associated with traditional craft, whilst also promoting the growth of new craft skills [20]. Luxury entrepreneurs can set an example in regards to developing production techniques that are environmentally sensitive, sustainable and ethical. Numerous luxury companies have extremely sophisticated strategies that enable them to encourage sustainability, which, once demonstrated to be effective, can be embraced by the economy [21]. Hence, sustainability is conveyed more to boost brand value than to pass on any information concerning the true effect of these activities [22]. Researchers are in agreement that the environmental and societal effects of consumption necessitate significant attention immediately and, moreover, that they need studying in more varied market settings [23].

3 Sustainability and Luxury Fashion in Saudi Arabia

In Saudi Arabia, the role of corporate social responsibility (CSR) is growing and having a positive impact on social culture [24]. Strong socio-cultural elements have already had a tremendous impact on the promotion of CSR initiatives [24]. Typi-cally, it is considered a social act or an act of humanitarianism, although, in the past, in Saudi Arabia, many organisations do not practice CSR as the bottom line with respect to their values [24]. The focus on CSR in Saudi Arabia has increased dramatically as the government endeavours to move from being a welfare state to a country that has a robust and burgeoning private sector [25]. Current issues, such as globalisation, have piqued the attention of countless organisations as regards human resources and organisational culture and consequently, they have begun to take their social responsibilities and response to business ethics much more seriously. In turn, these organisations are required to be more ethically and socially responsible with regard to people as well as others within their outside environment with the aim of meeting individual aspirations as businesses that are socially responsive and ethical

[26]. According to the study conducted by Abu Amara [27] on the Saudi industry and sustainability, social responsibility is more important than environmental sustainability. Saudi Arabia is a country that is rich in socially responsible business opportunities, together with several strong economic indicators as well as a cultural and religious history that is in keeping with the goals of its private sector's CSR work [28]. Furthermore, it should be mentioned that the Islamic tradition affords additional cultural or religious support to CSR initiatives in the business [28]. Business owners and entrepreneurs agree that CSR activities are important for them with respect to giving back to society [29]. Alajlouni [30] noted that significant reforms led by the Government of Saudi Arabia have moved from CSR in the governance hierarchy to the entrepreneurship level by way of engaging with businesses to achieve sustainable development, which is impacting fashion entrepreneurs in particular. In the past, there have been various challenges regarding the adoption of sustainable growth, which include government backing, stakeholder interest, sustainable resources alongside an absence of public awareness [10]. In Saudi Arabia, new fashion entrepreneurs try to use traditional methods in an established way so as to keep pace with modern developments, whilst taking into account preserving originality and valued roots. There is an emerging voice of young Arab designers who want to build a sustainable future for their consumers [31], thereby paving the way to combine traditional craftsmanship with twenty-first-century thinking of sustainability. These young Arab designers play a key role in the Saudi fashion industry, as they encourage new and established luxury entrepreneurs to produce and design clothing for contemporary life that is inspired by Saudi heritage. As mentioned previously, Saudi Arabia has witnessed this upsurge in creativity as regards design and fashion, together with a fresh emphasis on a way of life that is more sustainable and ethical and a developing youth culture that is encouraging, exciting, unprecedented and individual aesthetic. Saudi Arabia gives the impression that it is an inspiring trendy [32]. Moreover, research in Saudi Arabia has focused on popular innovation by collecting samples of traditional heritage, displaying and studying it to help the designers gain a distinct competitive advantage in the marketplace. The results of the study by Nazer and Mohammed [33] provided the following recommendations: Consider the heritage and pride of national identity in enriching fashion design with a contemporary vision of an economic community and the development of skills in accordance with the goals of the Saudi Vision 2030—an area that is focused on in this chapter.

4 Saudi Female Entrepreneur and Luxury Sustainable Fashion

On account of the economic, political and technological transformations in recent years in Saudi Arabia more and more females are now self-employed and own businesses. For those females, who wish to own their own business, these challenges have provided various economic opportunities [34]. Females in Saudi Arabia have gradually been taking a greater interest in fashion [35]. Traditional costumes in Saudi Arabia are an area in which females excel as they offer a variety of designs [33]. Princess Noura bint Faisal Al Saud (founder of the Saudi Fashion Week) stated:

> Fashion in Saudi Arabia is inherent to our story, our history and our heritage. In Saudi today, we are seeing the influences of fashion's past coming to the fore. More designers are referencing our heritage in contemporary ways and see the process of creating apparel as an opportunity to narrate the story of Saudi identity [36].

This quote highlights not only the increased economic importance of the Saudi fashion industry but also indicates its importance in terms of cultural traditions and heritage.

As previously stated, most female-run businesses operate within the fashion sector [37, 38]. The study conducted by Aboumoghli and Alabdallah [2] asserts that support from policymakers is likely to enable Saudi females to participate in entrepreneurial activities within the country's economy. An entrepreneur works in partnership with product weavers to produce merchandise that is fashionable and transform ethnic weaves into luxury pieces that last in terms of (1) their durability and quality and (2) their heritage and cultural meaning [39]. A fashion entrepreneur works diligently to merge traditional techniques with modern designs and creates (ideally) sustainable luxury products. Gardetti and Torres [40] assert that the relationship between the notion of entrepreneurship and sustainable luxury is incredibly close. Gardetti and Torres [40] believe that "sustainable luxury" comprises "craftsmanship, preserving the cultural heritage of different nations" (p. 55). Moreover, it is for "helping others to express their deepest values" [41]. This role of moving values and cultural heritage forward is particularly prominent with female fashion entrepreneurs, who are increasingly supporting Saudi Arabia's sustainable development plans not only through their designs but also through their economic importance as drivers for the economy [42]. To explain, not only do they create new employment opportunities for themselves among others, they have a different approach to organisational, management and business issues as well as having innovative ideas towards the exploitation of opportunities in the business context [43, 44].

5   Luxury Fashion and Cultural Heritage

Fashion alludes to the aesthetic, cultural and symbolic meanings that objects convey; specifically, the ways people make use of objects to exhibit their taste, social status, lifestyle as well as belonging to a community. Fashion represents symbolic meanings and shares a degree of mutual and social understanding (e.g. [45]). It can be described as a form of non-verbal communication that signifies symbolic and social consumption with the aim of managing identity [46]. Traditional costumes, crafts and combined experiences are certain types of heritage that require care so as to improve and be sustainable [47]. It should be mentioned that cultural heritage and handcrafted quality have high symbolic value; luxury items tended to be handed down from each generation to the next as heirlooms [40]. Thus, key characteristics of luxury are cultural values and authenticity, and more recently, the association with CSR and sustainability [48]. The creation of luxury that is culturally sustainable, subsequently, focuses on the aspects of manufacturing, cognition and attitude, besides identity, which is acknowledged as being different from those related to luxury that is conventional and sustainable [48]. Sustainable luxury supports the resurgence of unique art forms and presents a distinct stage for artisans to safeguard their local culture and incorporate it into both environmental as well as social issues by way of producing them [41]. Joy et al. [49] assert that immediate respect for handicrafts workers alongside the environment helps to encourage values that are in support of sustainability amongst young consumers of luxury fashion. Within the Saudi Arabian context, cultural sustainable products are associated with decorative styles and a variety of shapes, as well as traditional practices, such as embroidery or Al-Sadu [50]. The latter is a nomadic weaving technique primarily performed by women and still undertaken at present by weavers in the country [51]. This type of weaving demonstrates an abstract, symbolic language whose meanings have been communicated orally [51].

6 Methodology

This qualitative chapter draws on a rich data set, whereby 29 semi-structured interviews were undertaken with well-known Saudi female fashion entrepreneurs. Purposive sampling was employed to enlist Saudi female entrepreneurs, who have established their own sustainable luxury fashion businesses and have an interest in the cultural heritage and handicrafts and are over the age of 18. Appendix 1 provides a summary of the participants. All interviews were conducted online and recorded, after gaining consent. The average interview lasted 60 min. Interviews were carefully transcribed verbatim. All interviews were conducted in Arabic, translated into English and back again, to ensure no meaning was lost. To ensure the quality of the data collected, questions for the conversational semi-structured interviews were guided by key literature.

An inductive investigation was undertaken, adopting an interpretive approach with the aim of discovering additional reasons as well as to focus on greater awareness in relation to luxury entrepreneurs and sustainability. Data were analysed using the seven-step guide developed by Easterby-Smith et al. [52], which consists of the following: "familiarisation, reflection, open-coding, conceptualisation, focused re-coding, linking and re-evaluation". It is important to note that grounded analysis is open to discoveries within the data. To ensure coherence, clarity and continuity, the researchers examined the data independently. During this stage, they concentrated on the words and phrases that the interviewees used the most and studied them within their natural boundaries. The open-coding process enabled patterns to develop naturally. Concerning this study, one limitation observed may possibly be the interpretive nature of the research. Nevertheless, the findings obtained provide interesting information on luxury entrepreneurs. Additional attention was given to the cultural nuances of each of these personal conversations with the aim of ensuring that the translations are as accurate as possible.

7   Findings and Discussion

The interviews revealed the different ways entrepreneurs seek to achieve sustainability in their products through heritage. An overarching goal of the entrepreneurs interviewed for this research is to produce luxury garments with heritage touches in the pursuit of the Saudi Vision 2030. Although environmental sustainability relating to materials is not a key priority by our entrepreneurs, we found that a number of our interviewees have a greater awareness of what sustainability is and how to implement it in the future. The participants recognised the importance of international heritage and Saudi heritage; the latter has inspired some of these Saudi entrepreneurs. For instance, E12 states that "this vibrant cultural heritage has found expression in an extraordinary fashion collection, which combines the traditional fabric designs of our ancestors with contemporary trends". This view was echoed by another informant, E17 who says, "I have a love for heritage and I felt that the market was missing something specific and that the world needs culture as well as the quality of this heritage". As a result of their enormous symbolic value, luxury objects were typically passed down from one generation to the next. Thus, three key themes emerged relating to the significance and meaning of sustainability in luxury fashion: *1. Meanings and beliefs assigned to luxury sustainable fashion and Saudi Vision 2030; 2. Materials and design as symbols of luxury sustainable cultural heritage; 3. The interplay of heritage creativity in reinventing modern sustainable luxury fashion.*

7.1   Meanings and Beliefs Assigned to Luxury Sustainable Fashion and Saudi Vision 2030

As indicated in the literature review, improving the country's cultural landscape with the intention of illustrating Saudi Arabia's rich, national, artistic and creative heritage in fashion has been underlined as a key feature of the strategic objectives in the Saudi Vision 2030 [53]. With the aim of having a common understanding and enabling discussions to take place, gaining an awareness of what luxury sustainability in connection with culture means and an understanding of the goals of the Saudi Vision 2030 was considered crucial. Although participants typically demonstrated that they are preserving heritage, by including it (e.g. embroidery) in their fashion collections, E9 goes even further by stating that it is a "national duty" and part of sustainability that they seek to revive and preserve Saudi heritage from the principle of social responsibility. Simultaneously, each participant commits to the values associated with sustainable culture, although they are conveyed in different ways, representing the diverse approaches observed within the sustainable luxury fashion sector. Certain participants acknowledged that even though they do not understand the meaning of sustainability, they believe that it is essential to maintain Saudi heritage and follow the country by creating high-quality fashion that includes traditional production techniques. E19 states that "preserving heritage identity is one of my most important goals", seeing that it is linked to aspects of social responsibility. Murphy et al. [25] assert that Saudi entrepreneurs have high expectations in relation to corporations and social responsibility. Consequently, entrepreneurs responded to the future of the Saudi Vision 2030 and contribute to sustainable heritage by creating luxury fashion items in conjunction with heritage. E4 reported, "mixing between the heritage past and the dazzling present". Certain entrepreneurs seek to attain the Saudi Vision 2030 that displays Saudi heritage to the world by way of designing luxury garments. E16 alluded to the idea of making "garments that can be worn anywhere in the world" and believes that it could resonate well with women from different cultural backgrounds. Participants mentioned that this luxury sustainable product is a combination of Saudi heritage and Western tastes. However, some may be less conscious of sustainable development given that Saudi Arabia is a country that has a low level of awareness with respect to sustainable development [54].

On the other hand, some entrepreneurs are fully aware of the Saudi Vision 2030 and of what sustainability means in the fashion context, in connection with preserving heritage. E14 stated, "I am currently working on a project for Neom, which is part of Vision 2030, and I have designed a modern collection with heritage touches according to the specifications of the Vision that I hope will be accepted to be displayed soon". E14 specifically demonstrated a strong and personal interest in relation to the environment and manufactures sustainable clothing as regards to the environment and social responsibility in all their designs. Likewise, entrepreneurs could help to support the Saudi Vision 2030 by designing sustainable garments that combine modern design (luxury) with traditional craftsmanship. E15 believes that "with the recent changes, Vision 2030 fashion in Saudi Arabia has acquired a special identity, is a mixture of openness and originality, coupled with modernity in designs in a conservative character".

Several entrepreneurs contributed to the integration of modern aesthetic philosophy and traditional handicrafts into a Saudi brand that celebrates culture through timeless pieces. E28 stated that "the world doesn't need just quantities of clothes. We work in cooperation with craftswomen who create rare pieces with the aim of creating sustainable job opportunities for them over the coming years and preserving Saudi craftsmanship and heritage on the line of Vision 2030". To these craftswomen, this means designing reliable items comprising luxury modern heritage and employing complementary patterns so that over time, customers are encouraged to wear them in the long term. Having an established fashion market with marketing activities that are sustainable will enable long-term relations to be formed between traditional fashion market brands and customers [55].

Subsequently, it will enhance the concept of sustainability among consumers and Saudi society, which will result in the Saudi Vision 2030, which calls for fashion sustainability, being achieved. Some entrepreneurs had designed luxury sustainable garments with a heritage touch, which, to a degree, could encourage those people who have brand loyalty to be more aware. E14 asserted that "if a person owns a piece that bears the meaning of his or her history, I do not think it is easy to forfeit it... and I see that with my customers". Sustainable luxury seeks to return to the very heart of luxury. It implies thinking about a purchase, as well as the craftsmanship and decent materials, whilst making sure both the society as well as the environment are respected. Thus, it provides a chance to reclaim, protect and boost the cultural traditions of distinct areas and peoples [40].

7.2 Materials and Design as Symbols of Luxury Sustainable Cultural Heritage

Saudi Arabia is a multicultural and diverse country; therefore, people have different cultural and material expressions in their clothing, with an understanding that the origin of every single one is the creation of genuine services or pieces which are special, full of meaning and comprise anecdotes pertaining to its creator. Accordingly, a close relationship is established between the creator and their establishment throughout the entire process, which is then transmitted to the world, producing products that are durable and that consist of significant symbolic value.

Moreover, some entrepreneurs are using hand embroidery not only because of its beauty but also for modification and exploration based on their inspirations, seeing as embroidery performed by a machine cannot change embroidery based on the designer's inspirations. This timeless craft has been frequently invigorated by imaginative artists who push the boundaries in relation to its meaning and limits. One participant explained the tendency to use hand embroidery because she believes that hand embroidery helps creativity and changes the embroidery lines in a luxurious, modern way, that is currently appropriate for luxury fashion and producing rare pieces with a modern touch. E26 states that "I am trying to preserve a unique art by buying traditional female work and then turning it into jackets". Each traditional hand embroidery pattern has a different meaning-making and value construction [56]. It was necessary to design clothing and accessories that would not only include traditional crafts but would also make sense in the world of the modern consumer [56]. As E8 commented "no matter how advanced embroidery is using modern machines, hand embroidery is indispensable for us, because it is authentic, solid and has respect for our heritage. It is also close to the heart. I have pearls on my designs. That is why I see that it is impossible for any designer to dispense with hand embroidery. This results in him being careful that the craftsmanship attached to it does not disappear. Also, no modern machinery can match the imagination of fashion designers".

Furthermore, it is important to note that luxury, sustainability as well as ethics introduce dialogue that is constant and unchangeable in relation to the modification of the fashion system. Several meanings are available with respect to endorsing styles which are sustainable. These include using resources in a more intelligent way to reducing the effect on the environment, paying workers a fair wage and ensuring that they work in a safe environment, preserving traditional customs to promoting studies on resources that are biocompatible along with circular production systems where excess is reduced with the aim of avoiding goods ending up in landfill sites [57]. Regarding process optimisation, even from the perspective of traditional crafts-manship, innovation is essential to ensure that the excess remaining from resources and materials is reduced. E17 explained when she visited the global village in Dubai "there is an amount of heritage in the Pakistan, India, and Afghanistan section of the exhibition, which is impressive in the amount of heritage that exists... I asked myself. Why is it cut off everywhere? Is someone who has done something from it? I took it and my first design appeared". On the other hand, as part of social responsibility E27 indicated "try to help the local female handicrafts by sustainable employment for the many years as well as preserve the cultural heritage craftsmanship of Saudi Arabia".

7.3 The Interplay of Heritage Creativity in Reinventing Modern Sustainable Luxury Fashion

Entrepreneurs recognised the uniqueness of the rich international and local historical context that was used to produce luxury sustainable garments that were fashionable, of excellent quality and consisted of products that were durable as a part of their adherence to social responsibility. Saudi female entrepreneurs work in partnership with hand product weavers to produce contemporary products and remodel ethnic weaves into luxury items that last. E17 explained, "it became one of my goals to revive world heritage... Tunisian and Moroccan...Saudi...Indian, Pakistan, Afghanistan...I do not want people to see the piece and say I cannot wear these garments because they are heritage. I want the piece to be and live in the closet for a period of time that I care about the sustainability of the product and the sustainability of the heritage". E17 works diligently to merge world traditional techniques with modern designs and create sustainable luxury products. These days Saudi female entrepreneurs are incorporating local cultural heritage in their luxury collections as a part of their personal and social identity. Several of their collections were inspired by ancient traditional costumes and different tribal identities from a variety of regions in Saudi Arabia. E16 said, "one of the looks that I created from this collection is the white T-shirt, paired with a fashion-inspired tulle wrap skirt worn by the traditional man in the Asir region of the Kingdom..., which is decorated with hand-embroidery that forms geometric patterns with a gold and black texture on the tulle fabric". Also, "I have always given my collections names that are related to Saudi heritage. Female entrepreneurs have not only used designs from traditional female garments but have incorporated traditional men's fashion with touches of feminine, luxury designs that could help to make men's and women's fashion sustainable".

Currently, some Saudi female entrepreneurs have managed to incorporate local cultural heritage in luxury collections. Al-Sadu, which is recognised by the UNESCO, is the perfect sustainable material in regards to the development of craft-based luxury. Participants disclosed that meanings concerning the dress were passed down through intergenerational rites of passage. E23 commented, "my collection 'Honour' is inspired by Saudi heritage as I want to introduce the world to our wonderful heritage..(She used) hand embroidery—"Al Sadu"—from North (Saudi)". On the same line of thought, E28 said, "one of the models sported a giant woven scarf made from a traditional fabric called "Al-sadu"", the parka jacket was lined with red and white. Fashion heritage represents a strong and specific asset for generating creativity. It is important to state that fashion heritage is, at the same time, a physical accumulation of models and designs, as well as the cultural legacy of a particular style. It is incredibly useful for the reason that designers can reconsider previous styles and collections and produce them as contemporary designs. Regarding its progression, fashion develops the dynamic of permanence and change.

Finally, the Saudi Vision 2030 stresses that women in Saudi Arabia have a vital role to play in regard to the nation's development strategy [58]. The Saudi 2030 Vision emphasises femininity, particularly being young, energetic, entrepreneurial and successful. Moreover, Saudi 2030 Vision emphasises femininity, particularly being young, energetic, entrepreneurial and successful. Moreover, Saudi Vision 2030 aims to promote the position of women regarding the impact they are having in relation to the sustainable development plan. This Vision has resulted in a notable improvement in Saudi Arabia. E20 claims that the "Vision 2030 has embraced and supported our fashion sector and highlighted the role of women in the past and present". It is worth noting that promoting the participation of women in the fashion industry is an important part of the Saudi Vision 2030 and the national transformation programme. Additionally, females are not only entrepreneurs but also seen as creative entrepreneurs as they participate in cultural and economic values.

8 Conclusions

This study seeks to highlight the role of the Saudi Vision 2030 in the fashion of the future in an attempt to develop sustainable ideas for fashion luxury entrepreneurs. Specifically, it should be mentioned that this chapter is embedded in the well-established idea of traditional fashion. The initial findings highlight that the female entrepreneurs chosen for this sample are currently transitioning from creating traditional garments towards those that incorporate modern-day issues of sustainability. The Saudi Vision 2030 has provided these entrepreneurs with an opportunity to develop businesses that focus on environmental, social and economic aspects by following traditional production processes. Although the Saudi market has various traditional garments, it must also develop a more modern style that could be attractive to "outsiders"—those not of Saudi origin but interested in the culture. The results of the study are significant as luxury female entrepreneurship has an important part to play in the growth of the concept of luxury fashion sustainability that relates to cultural heritage in Saudi. In light of recent economic reforms, this study provides valuable insight into female entrepreneurship in the Kingdom of Saudi Arabia. Additionally, this study helps to understand the importance of Saudi heritage in enriching luxury sustainable fashion design with a contemporary artistic vision, which is in line with the goals of Vision 2030 that could develop traditional manual skills and stimulate traditional crafts in Saudi Arabia to be more sustainable. Furthermore, the survival of cultural heritage and a return to the business's emblematic and social assets is significant, especially when looking at the entrepreneurs and their shareholders and how they work togetehr to fulfil the Saudi Vision 2030. Current research on luxury sustainable fashion with a focus on cultural heritage in Saudi Arabia was evaluated in conjunction with new existing legal and governmental standards. These collective energies indicate change and also what is known as "the epiphanic sustainable fast fashion epoch—a new ethical fashion" directive. However, the current potential barriers identified relate to the reach of the garments and their designs, based on the fact that Saudi Arabia remains a conservative culture even though it is gradually opening up. A limitation could be the focus on the female sample of entrepreneurs; however, female entrepreneurs in Saudi Arabia have received increased interest and there is an increased number of female fashion entrepreneurs in Saudi. Luxury entrepreneurs believe that they make a positive contribution to society. The examples provided indicate that luxury products are not only concerned with promoting their reputation but also in demonstrating moral and altruistic values by way of several sustainable practices.

Appendix 1—Summary of Participants

No	Industry segment
E1	Designer Luxury Womenswear (Evening Apparel)
E2	Designer Luxury Womenswear (Evening Apparel and Casual)
E3	Designer Luxury Womenswear (Ready-to-Wear)
E4	Designer Luxury Womenswear (Evening Apparel)
E5	Designer Luxury Womenswear (Evening Apparel)
E6	Designer Luxury Womenswear r (Evening Apparel and Casual)
E7	Designer Luxury Womenswear (Ready-to-Wear) (Evening Apparel)
E8	Designer Luxury Womenswear (Children, Casual, Weddings and Evening Apparel)
E 9	Designer Luxury Womenswear (Ready-to-Wear) (Abaya)
E10	Designer Luxury Womenswear (Ready-to-Wear)
E11	Designer Luxury Womenswear (Evening Apparel) (Ready-to-Wear)
E12	Designer Womenswear Sustainable Fashion Ethical Luxury (Ready-to-Wear)
E13	Designer Luxury Womenswear (Ready-to-Wear)
E14	Designer Womenswear Sustainable Fashion, Ethical Luxury (Ready-to-Wear)
E15	Designer Luxury Womenswear Haute Couture (Ready-to-Wear)

(continued)

(continued)

No	Industry segment
E16	Designer Luxury Womenswear
E17	Designer Womenswear Sustainable Fashion, Ethical Luxury (Modern Folk Garment)
E18	Designer Womenswear Fashion, Luxury (Ready-to-Wear)
E19	Designer Womenswear Fashion, Ethical Luxury (Modern National Garment (Abaya)
E20	Designer Womenswear Fashion Luxury (Bags, Accessories, Apparel)
E21	Designer Womenswear Fashion, Ethical Luxury (Modern National Garment (Abaya)
E22	Designer Womenswear Fashion, Luxury (Bags, Accessories, Apparel)
E23	Designer Womenswear Fashion Ethical Luxury
E24	Designer Womenswear Fashion
E25	Designer Womenswear Sustainable Fashion Ethical Luxury
E26	Designer Womenswear Sustainable Fashion Ethical Luxury
E27	Designer Womenswear Sustainable Fashion
E28	Designer Womenswear Fashion Luxury (Bags, Accessories, Apparel)
E29	Designer Womenswear Fashion Luxury (Bags, Accessories, Apparel)

References

1. Al Othman FA, Sohaib O (2016) Enhancing innovative capability and sustainability of Saudi firms. Sustainability 8(12):1229
2. Aboumoghli AA, Alabdallah GM (2019) A systematic review of women entrepreneurs opportunities and challenges in Saudi Arabia. J Entrep Educ 22(6):1–14
3. Saudi Vision 2030 (2020) Vision 2030 Kingdom of Saudi Arabia. http://vision2030.gov.sa/en. Last Accessed 15 Oct 2020
4. Mostapha AHF, L, (2020) Wassem fashion district. J Crit Rev 7(8):535–537
5. Rana D, Alayed R, S, (2018) Green business sustainability within Saudi vision. Int J Adv Study Res Work 1(9):2581–5997
6. Long DE (2005) Culture and customs of Saudi Arabia cuclture and customs of the middle east. Westport CT Greenwood Press
7. Feda LAG (2021) Governing standards for the simulation of traditional clothing in Saudi Arabia. Int Des J 11(1):281–298
8. Tawfiq W, Marcketti S (2017) Meaning and symbolism in bridal costumes in western Saudi Arabia. Cloth Text Res J 35(3):215–230
9. Shafee WH, Feda LA (2019) Innovative solutions for traditional Saudi Arabian costumes using TRIZ principles. Int J Syst Innov 5:(3) 47–61
10. Dekhili S, Achabou MA, Alharbi F (2019) Could sustainability improve the promotion of luxury products? Eur Bus Rev 31(4):488–511
11. The Ministry of Saudi Culture (2019) Our cultural vision for the Kingdom of Saudi Arabia. https://www.moc.gov.sa/sites/default/files/2019-04/Cultural_Vision_EN_0.pdf. Last Accessed 14 Dec 2020
12. Davies I, Oates CJ, Tynan C, Carrigan M, Casey K, Heath T, Henninger CE, Lichrou M, McDonagh P, McDonald S, McKechnie S, McLeay FO, Malley L, Wells V (2020) Seeking sustainable futures in marketing and consumer research. Eur J Mark 54(11):2911–2939

13. Henninger CE, Alevizou PJ, Goworek H, Ryding D (eds) (2017) Sustainability in Fashion: a cradle to upcycle approach .Springer
14. Henninger CE, Alevizou PJ, Oates CJ (2016) What is sustainable fashion? J Fash Mark Manag 20(4):400–416
15. Fashion Futures (2019) Why fashion futures 2030. Fashion futures2030.com. https://fashionfutures.com/en/home. Last Accessed 13 Nov 2020
16. Li J, Leonas KK (2019) Trends of sustainable development among the luxury industry. In: Sustainable luxury. Springer, Singapore, pp 107–126
17. Gardetti MA, Muthu SS (eds) (2019) Sustainable luxury: cases on circular economy and entrepreneurship. Springer
18. Gardetti MA (2011) Sustainable luxury in Latin America. In: Conference dictated at the seminar sustainable luxury and design, Instituto de Empresa (Business School). Madrid, Spain
19. Athwal N, Wells V, Carrigan K, M and Henninger C E, (2019) Sustainable luxury marketing: a synthesis and research agenda. Int J Manag Rev 21(4):405–426
20. Roberts J (2019) Luxury international business: a critical review and agenda for research. Crit Perspect Int Bus 15(2):219–238
21. Armitage S, Hou W, Sarkar S, Talaulicar T (2017) Editorial: corporate governance challenges in emerging economies. Corp Gov 25(3):148–154. https://doi.org/10.1111/corg.12209
22. Kunz J, May S, Schmidt HJ (2020) Sustainable luxury: current status and perspectives for future research. Bus Reshttps://doi.org/10.1007/s40685-020-00111-3
23. Osburg VS, Davies I, Yoganathan V, McLeay F (2020) Perspectives, ppportunities and tensions in ethical and sustainable luxury: introduction to the thematic symposium. J Bus Ethics https://doi.org/10.1007/s10551-020-04487-4
24. Sharmaa RB (2020). Identifying and ranking the key performance drivers of corporate social responsibility in Saudi Arabia. Int J Innov Creat Change 10(10):483–494
25. Murphy MJ, MacDonald JB, Antoine GE, Smolarski JM (2019) Exploring muslim attitudes towards corporate social responsibility are Saudi business students different? J Bus Ethics 154(4):1103–1118. https://doi.org/10.1007/s10551-016-3383-4
26. Alharbi JM, Alharbi SM (2019) Business ethics social responsibility and competitive advantage: the Saudi case. J Bus Econ Policy. https://doi.org/10.30845/jbep.v6n1p4
27. Abu Amara M (2019) Disclosure of environmental responsibility in Saudi joint stock companies. Glob J Econ Bus 7(3):286–302
28. Almalkawi HAN, Javaid S (2018) Corporate social responsibility and financial performance in Saudi Arabia: evidence from Zakat contribution.Managerial finance 44(6):648–664
29. Alsabban N, Al sabban Y, Rahatullah MK (2014) Exploring corporate social responsibility policies in family owned businesses of Saudi Arabia. Int J Res Stud Manag 3(2):51–58.
30. Alajlouni MM (2020) Governance and social responsibility practices and activities at Saudi business schools. Governance 5(2):26–35
31. Tewari B (2019) These millennial Arab designers are at the forefront of sustainability .Harpersbazaararabia. https://En.Vogue.Me/Fashion/Designers/Millennial-Arab-Designers-Sustainability/. Last Accessed 14 Dec 2020
32. Khan R (2020) Exclusive: Saudi label Yasmina Q is the new sustainable fashion line you need to know. Harper's Bazaar Arabia. https://En.Vogue.Me/Fashion/Sustainable-Saudi-Label-Yasmina-Q-Launches/. Last Accessed:24 Dec 2020
33. Mohammed NEAAH, N A R, (2018) The aesthetic values of bead handicrafts inspired by Saudi heritage in Taif Governorate ornamentations utilized in design on the dress form. Int Des J 8(2):139–150
34. Basaffar AA, Niehm LS, Bosselman R (2018) Saudi Arabian women in entrepreneurship: challenges,opportunities and potential. J Develop Entrep 23(02). https://doi.org/10.1142/S1084946718500139
35. Upadhyay R (2016) Saudi women designers highlighted at event in Jeddah. https://wwd.com/fashion-news/fashion-scoops/vogue-italia-jeddah-10421492/. Last Accessed 13 Jan 2021
36. Phillips O (2020) Her highness princess Noura bint Faisal Al Saud on her mission to see Saudi fashion thrive on a global scale.harpersbazaararabia. https://www.harpersbazaararabia.com/

featured-news/her-highness-princess-noura-bint-faisal-al-saud-interview-harpers-bazaar-ara bia-saudi. Last Accessed 14 Dec 2020

37. Almunajjed M (2010) Women's employment in Saudi Arabia: a major challenge. Booz and Company
38. Esmail H (2018) Economic growth of Saudi Arabia between present and future according to 2030 vision. Asian Soc Sci 14(12):1911–2025
39. Jain S, Mishra S (2019) Sadhu—on the pathway of luxury sustainable circular value models. In: Sustainable luxury. Springer
40. Amarilla R Gardetti MA, Gabriel M (2020) Sustainable luxury, craftsmanship and Vicuna Poncho. In: Sustainable luxury and craftsmanship. Springer, Singapore
41. Gardetti MA, Torres AL (2013) Entrepreneurship innovation and luxury. J Corp Citizsh 52:55–75
42. Alkhateeb AMMZ, T T Y Mahmood H Abdallah M A Z qaralleh T J O T, (2020) The role of the academic and political empowerment of women in economic, social and managerial empowerment: the case of Saudi Arabia. Economies 8(2):45
43. Aldajani and Marlow (2013) Empowerment and entrepreneurship: a theoretical framework. Int J Entrep Behav Res 19(5):503–524
44. Ramadani V, Gërguri S, Dana LP, Tašaminova T (2013) Women entrepreneurs in the Republic of Macedonia: waiting for directions. Int J Entrep Small Bus 19(1):95–121
45. Alevizou PJ, Henninger CE, Stokoe J, Cheng R (2021) The hoarder, the oniomaniac and the fashionista in me: a life histories perspective on self-concept and consumption practices. J Consum Behav Ahead Print
46. Ostberg J (2012) Masculinity and fashion. In: Gender, culture, and consumer behavior. Otnes CC, Zayer LT (eds) Routledge. New York
47. Shafee WH (2017) The effect of simulating traditional costume images in the development of innovative capabilities for contemporary fashion design. Int Des J 7(4):307–314
48. Ranfagni S, Guercini S (2018) The face of culturally sustainable luxury: some emerging traits from a case study. In: Sustainable luxury, entrepreneurship, and innovation. Springer, Singapore
49. Joy A, Sherry Jr JF, Venkatesh A, Wang J, Chan R (2012) Fast fashion, sustainability and the ethical appeal of luxury brands. Fash Theory 16(3):273–295
50. Alajaji TN (2019) Traditional embroidery as a method of decoration in Najd desert tribes. J Text Sci Fash Tech (JTSFT) 4(3):000587
51. Alogayyel R, Oskay C (2020) Al-Sadu weaving: significance and circulation in the Arabian Gulf. In: All things Arabia.brill
52. Easterby-Smith M, Thorpe R, Jackson PR, Jaspersen LJ (2018) Management and business research. Sage
53. Mohammadian-Molina RM (2020) Fashion tech: how Saudi Arabia could take the lead. In: This untapped sector. https://www.entrepreneur.com/article/347442. Last Accessed 13 Jan 2021
54. Alsabban L, Issa T (2020) Sustainability awareness in Saudi Arabia. In: Sustainability awareness and green information technologies. Springer, Cham
55. Jung J, Kim S, Kim J, K H, (2020) Sustainable marketing activities of traditional fashion market and brand loyalty. J Bus Res 120:294–301
56. Khaire M (2019) Entrepreneurship by design: the construction of meanings and markets for cultural craft goods. Innovation 21(1):13–32
57. Cappellieri A, Tenuta L, Testa S (2020) Jewellery between product and experience: luxury in the twenty-first century. In: Sustainable luxury and craftsmanship. Springer, Singapore
58. Alemam R (2020) Saudi women rising up in business in line with Vision 2030. https://www.worldbank.org/en/news/feature/2020/03/11/saudi-women-rising-up-in-bus iness-in-line-with-vision-2030. Last Accessed 13 Jan 2021

A Sustainable Alternative for the Woven Fabrics: "Traditional Buldan Handwoven Fabrics"

Gizem Karakan Günaydın⬡ and Ozan Avinc⬡

Abstract As the world population and the consumer demands have increased, the promising ways for a sustainable textile production are much more considered more than ever. Handwoven textiles may be preferred as a good alternative for reducing environmental impact during the textile production. A return to traditional manufacturing will not only provide a sustainable process with utilizing less energy sources but also ensures that cultural heritage is passed on to future generations creating new employment opportunities. Textile and weaving have vital factors on the economical prosperity of cities and countries. Aegean Region in Turkey has been a major area for cotton cultivation and weaving for centuries. "Buldan" town of Denizli city has been known as the weaving center where the famous "Buldan cloth" is manufactured. It was officially proved in the early historical documents that high amount of the woven fabric needs of Royal Palace of Ottoman was provided from Buldan town. Although Buldan fabrics were generally produced for underwear and home textile first, today they can even be utilized for outdoor garments owing to fabrics' satisfying comfort properties. Those fabrics have also high elasticity with their high twisted yarns. People living in Buldan started textile production many years ago by producing different woven fabrics in their rooms or in the basements of their homes. As a result, this mode of production has made the current Buldan weaving industry not just a business, but a lifestyle in Buldan. Therefore, textile started many years ago in the houses of Buldan which makes weaving sector not just a business but also a way of life and tradition transfer tool in Buldan. Famous traditional cloth "Alaca" cotton colored striped garment has been woven since nineteenth century by the town people. Large number of handkerchiefs, home textile products such as bed coverings, towels, vivid violet silk (pesthemal) to be wrapped around the body, and colorful head coverings are also noteworthy to be mentioned among the wide variety of Buldan textile products. Traditional handwoven products are still produced and sold in Buldan. However, in addition to handwoven textile products, with the widespread industrial

G. K. Günaydın (✉)
Buldan Vocational School, Pamukkale University, Buldan, Denizli, Turkey
e-mail: ggunaydin@pau.edu.tr

O. Avinc
Textile Engineering Department, Pamukkale University, Denizli 20160, Turkey

© The Author(s), under exclusive license to Springer Nature Singapore Pte Ltd. 2021　　87
M. Á Gardetti and S. S. Muthu (eds.), *Handloom Sustainability and Culture*,
Sustainable Textiles: Production, Processing, Manufacturing & Chemistry,
https://doi.org/10.1007/978-981-16-5967-6_5

development in the town, semi-automatic and advanced automatic weaving looms also began to be used. So, as the industrial development has spread in the town, woven advanced automated looms have been started to be used widely which resulted with product exporting to many different countries in piece/part and metered weave form. Thanks to the sustainable and environmentally friendly production effect of handwoven products and the mass production potential of semi-automatic and automatic weaving machines, the production of Buldan handwoven and machine-woven fabrics and the export of textile products produced from these fabrics [in the form of finished product such as clothing (t-shirt, shirt, short, trousers, dress, scarf, etc.), home textiles (towels, pesthemal, table linen, bed coverings, curtains, etc.), piece goods and over-the-counter-fabrics] to many different countries is increasing day by day. Buldan weaves are special for their revealing a unique fabric structure with traditional designs, colors. These fabrics are also popular for providing comfortable clothes made of natural fibers such as cotton, wool, linen (flax), and silk. Hence, they are good alternatives for the today's sustainable textile production with less chemical usage. In addition to the use of natural and sustainable textile fibers in these products, the sustainable effect of hand weaving also increases the consumption and preference of Buldan handwoven products. High twisted yarns are utilized in the traditional "Buldan twisted clothes" where plain weaves are generally preferred. High twist level provides the high thermal insulation where those fabrics do not contact the skin beside with the ease of movement. Buldan fabrics made of 100% cotton with high sweat absorption has been very attractive for the healthy product consumers. The purpose of this chapter is to present a general aspect and information about traditional Buldan weaves, Buldan handwoven fabrics, woven different textile products, the utilized handlooms in the town, the importance of weaving on the economical structure in Buldan district as well as some traditional hand weaving centers that still exist today in Buldan.

Keywords Buldan handloom fabrics · Buldan weaves · Buldan handwoven · Hand weaving · Buldan · Traditional weaving · Sustainable textile production · Culture · Handloom

1 Introduction

Sustainability is a complex issue and sustainability aims to counterpose the requirements of today without compromising the capability to counterpose the necessities of future generations to reach sustainable world. The sustainability is much more considered in the recent years as the world is polluted due to the high rate of industrialization. Environmental pollution increment resulted with the requirement of taking some precautions in all areas. As is known, there are high amounts of chemicals, water and energy consumptions in the textile processes. Thus, new sustainable production alternatives began to replace those that are not environmentally friendly, and this sustainable production trend is enhancing day by day and exhibits increasing development and growth [1–17]. In addition, the use of natural fibers, biodegradable fibers, and recycled fibers in the textile industry makes great contributions to the sustainability issue in textiles [18–33]. A new approach for marketing as "sustainable fashion" started to be widely adopted where ecological impacts of the textile product should be declared to the consumers. Most sustainability initiatives in the textile industry, but not all, focus on environmental sustainability. Therefore, when examined in the field of textile industry, it is observed that social and economic sustainability concepts may sometimes remain in the background. Hand-made productions may be accepted as an important way of sustainable textile production in many aspects such as utilizing less world energy sources with the continuity of employment. Hence, the social sustainability and the environmental sustainability can be mentioned both in hand-made production which allows traditions and customs to be passed down from generation to generation [34–42].

Textile has always been a part of the Turkey's culture and has been as the main part of the country's economy. Textile and apparel sectors are accepted as the most important sectors of the Turkey's economy. Looking at the world's leading home textile exporting countries; it is seen that countries such as China, India and Pakistan have a large share of this cake. One of the world's leading exporters of home textiles is Turkey. The city of Denizli plays a major role in the home textile exports of Turkey. Approximately 57% of the towel, bathrobe, and bed linen exports is provided from Denizli [43]. Denizli has been one of the important centers of hand weaving in the world for many years. Handwoven products are especially preferable for being niche designs. Hand weaving production has many factors to be considered in terms of sustainability. Generally, electricity is provided from petroleum or natural gas. Hand weaving may be accepted as an important sustainable textile process since it is not related with utilization of electricity power moreover it requires human power. Therefore, they do not harm the environment, in that way, when compared with other industrial processes.

Considering that robots will be more involved in the sectors in the coming industry 4.0 era, it is thought that hand weaving will contribute to the sustainability of human workforce. Handwoven fabrics with the traditional motifs provide an opportunity for the sustainability of the traditions of the society and to be passed on to the next generation for many years. Craftsmen teaching weaving to the next generation may be an example for the sustainability of knowledge and technique. Figure 1 indicates a craftsman weaving on the handloom. This craftsman has been weaving with this handloom since he was 16 years old (Fig. 1). The handloom in the photo (Fig. 1) is inherited from his father. As seen in this example, hand weaving is a craft passed from father to son in Buldan.

Most of the handwoven products may be colored by using natural products with the natural dyeing processes [44–46]. Coloring the yarns that make up handwoven Buldan products with natural resources makes an extra contribution to the sustainability of the textile. For example, Fig. 2 shows an example of handwoven products made from naturally dyed yarns with pinecones. The eco-friendly dyeing of Buldan fabrics may be also conducted with other regional plants such as sumac, acorn, walnut, onion, pomegranate peel, etc.

Fig. 1 A craftsman (Mr. Habip PEKÖZ) hand weaving using the handloom in Buldan district of Denizli

Fig. 2 Buldan fabric made from yarns dyed with pinecones

There are many textile products such as towels, bathrobes, sheets, curtains, and bed linens, exported to all over the world today, are produced in Denizli. This development is thought to be attributed to the establishment of the first yarn mills in the 1930s. These yarn factories were established to meet the yarn requirement of the small industry of home and workshop type that weaves raw cloth, which is very common in Denizli and its surrounding towns [43, 47]. As we classify the industrial establishments in Denizli, we see that most of these industrial establishments are gathered in the weaving, clothing, and leather sector, which ranges from 45 to 50%. This development accelerated especially with the establishment of ginning factories. With the establishment of modern industrial facilities, ginning factories were first started to be established in Denizli. It was observed that yarn factories were established as a large facility in Denizli after the cotton gin factories. With the opening of the yarn spinning factories, many different textile production sites working in other textile fields, especially weaving, began to open. Denizli province, including the Buldan district, has always been as one of the major areas for cotton cultivation providing good raw materials for the craftsmen of the district. The existence of the mineral springs led the area become an attractive "Spa" Centre since Roman times. One of the UNESCO World Heritage sites in the area "Pamukkale" welcomes the travelers with these healing waters and is worth to be visited [48]. For this reason, traditional hand weaving, historical, and natural beauties go hand in hand, making this city a place to be seen and discovered.

This chapter describes a comprehensive review of traditional Buldan handwoven fabrics in detail. Firstly, Buldan town of Denizli is introduced including its textile history starting from Hellenistic and Roman Empire periods. The information related the traditional Buldan plain fabric "Bükülü" utilized in Ottoman times is given. After a general overview of weaving of Buldan town, textile weaving and its importance in Buldan is emphasized. In the next part, traditional Buldan fabrics such as Buldan twisted cloth "Buldan Bükülüsü" and pesthemal are given in detail. In addition to these products, some more traditional products in Buldan such as handkerchiefs, fine shirting fabrics, white linings, and silk fabrics, home textile products such as bed sheets, tablecloth, pillow cover, cushion covers are also mentioned. Following the part related to traditional Buldan products, weaving machines utilized in Buldan starting from the primitive hand looms are described in the next part. Some famous expertised craftsman weavers in Buldan are also introduced. Moving on, next comes the "Embroidery in Buldan clothes" where the most confronted embroideries in Buldan clothes are discussed. Some of the embroidery techniques utilized in Buldan clothes are given with the detail information of thread types used for embroideries. Some special embroidery techniques in Buldan such as Suzani are displayed with the original images. The visual images of embroided tablecloths, coffee tablecloth, table napkins, etc., made from traditional Buldan fabric are also exhibited in this chapter. Moreover, some previous studies in the literature about Buldan cloth are also discussed. Afterwards, the information related to the examples of Local Workshop Centres for Hand weaving in Buldan" is given. For example, some local workshop centers such as "Evliyazedeler Konağı" and "Belsam (Buldan Handicraft Center)" are introduced in this final section of this chapter.

2 Buldan Town and Its History

Buldan town of Denizli is located in the northwest of Denizli city center and it is about 45 km away from Denizli city center. Its height above sea level is 690 m. Since Hellenistic and Roman Empire periods, Buldan region is one of the most important textile centers being on the trade road of "Büyük Menderes" where the trade from the inner part of Anatolia to the ports of Ephesus took place. The history of the region is very old. The region which was invaded by the Luwians and then added to imperial lands by the Hittites (1650–1200 BC) was added by the Phrygians and Lydians in the Iron Age (twelfth–seventh centuries BC) BC. In the sixth century, Persians included Denizli and Anatolia to their surroundings and expanded their borders. B.C. In the fourth century, Anatolia was included to the Macedonia by Alexander the Great. In the Hellenistic period that started with Alexander the Great, Seleucids and Pergamon Kingdom was active in the region. BC at 133, the lands were taken over by the Romans due to will of Attalos. Buldan and its region, which was included in the Byzantine borders after it was divided into two as East and West in 395, remained under Byzantine rule until the arrival of the Seljuk Turks. Accession of Buldan region to Ottoman lands coincides with 1429. During the War of Turkish

Independence, the settlement, which was occupied by the Greeks, was saved from the occupation on September 4, 1922. And Buldan to Denizli after Denizli became a province in 1923 with the Turkish Republic has become a connected district [49].

The hand weaving tradition in the district goes back to ancient times. It has been learned from the historical records that the Osmanoğulları supplied fabrics from Buldan through the Germiyanoğulları before they settled in Bursa. After the establishment of the Ottoman Empire, a part of the textile requirement of the Palace was provided by Buldan. Ottoman documents about the seventeenth century reveal that Buldan had been an important textile production centers for a long time. In Ottoman records, a Buldan cloth "Bükülü" was mentioned which has plain weave with 14 threads/cm warp-weft density. The fabrics usually made from silk or cotton fiber yarns reflecting the cultural identity of the region were pointed out where the traditional woven motifs were added [50]. Until the 1650, the cotton woven fabrics from Buldan was sent to town of Tire (Izmir) to be dyed. After this time, part of the operation also commenced to be handled locally. Even today, a sign at the entrance of Buldan town says: "Buldan greets the visitors with the pride of weaving the robe of I. Beyazid since he married Hafsa Hatun, daughter of Aydınid Isa Bey, in 1390." The presence of the first century AD Roman foundation in Tripolis (Phrygia) is thought to be the reason for the emergence of the weaving industry in this region. [51]. After a general overview of Buldan town and its history in this part, next part evaluates the importance of weaving and the economical structure in Buldan.

3 The Importance of Weaving and the Economical Structure in Buldan

Textile weaving is one of the important income sources of the district. Indeed, weaving has been one of the main livelihoods of the people of the region for centuries. Hand weaving in the district tradition goes back a long time. When Buldan's social situation is considered, it is observed that trade is the main factor that accelerates the development of the town. It is very interesting that there are many home–workshop associations in Buldan where the whole family including the children becomes the part of the weaving task. Hence it is understood that textile in Buldan is a lifestyle and business. Family businesses in Buldan meet their resources themselves for their workshops. They do not employ any workers outside their family, or they do not work for any other employments. The basic characteristics of weaving as an economic activity can be listed as follows labor-intensive technology, organization of production around the family, shortage of capital, irregular working hours, no wage relationship, small stocks and fixing the price by negotiation [52]. The weaving industry in Buldan may be classified into 4 categories as home industry, small industry, collective work in workshops, factory-type business.

Some weaving tools and equipment were renewed according to the conditions of that time. Since 1952, besides hand looms, motorized looms have been used in and around the district. Some handlooms had begun to be replaced by motorized looms. According to the records of Denizli Chamber of Commerce, there were 2450 hand looms in Buldan, where only 165 motorized weaving looms were located in 1960. It is understood that the most important activity was carried out via hand loom weaving in those times. This may reveal that there was no significant economic class differentiation among the small producer population. The number of handlooms in the town dramatically decreased to 1500 in 1984. There are around 20 hand weaving looms that are actively used today in Buldan [53].

Buldan hand weavers state that the Buldan cloth is an heirloom of their father and they say that they will try to continue legitimizing their profession. Handwoven of Buldan cloth seems to be hobby, love more than a trade for them. However, there is a big risk of disappearing of Buldan handwoven fabrics due to the high costs and difficulty of manufacturing. Although the weaving business started with the handlooms in the district, nowadays most of the woven fabric production continues with advanced automatic looms. While the textile products such as caftan, shirt, wedding dress, robe, and loincloth were handwoven here in the past, the products such as blouse, shirt, dress, scarf, tie, and loincloth are traditionally produced and sold today according to the age's clothing style though with the similar originality and quality. However, it is important to remark that hand weaving is not completely disappeared in Buldan, as mentioned earlier, hand weaving is already carried out in Buldan and different hand weaving products are produced and sold. Those traditional products and their structure were mentioned in the following parts of this chapter.

4 Traditional Buldan Fabrics

Plain woven fabrics produced with the handlooms in Buldan are called as "Buldan cloth" or "Buldan fabric." Buldan fabrics have natural appearances and basic textile features. The products made from Buldan cloth are generally classified into two groups because of the production amount on the loom: Piece production and metered production. Products such as pesthemal (a type of loin cloth or waist cloth and can be spelled as peshtamal, pestamal, or pestemal and spelled as peştemal in turkish language), tablecloth, towel, scarf are piece products, while products such as curtain, bathrobe, bedspread, sheet are classified under the meter-produced fabrics. The most traditional Buldan fabric "Buldan Twisted Cloth" and "the other traditional Buldan products" were examined in detail as the subtitles of this part [54].

4.1 Buldan Twisted Cloth

One of the most important traditional textile products produced by hand weaving in Buldan is known as Buldan twisted locally named as "Buldan Büklüsü" (Fig. 3). This weaving can be carried out using all kinds of classical natural materials such

Fig. 3 Example of Buldan Bükülüsü

as wool, linen (flax), cotton and silk. The most popular of its kind is made with silk threads. In Ottoman Times, twisted Buldan silk fabrics were generally utilized by the palace people, the rulers and the rich people. The public could not afford to buy silk twisted fabrics. For this reason, striped weaving samples made by adding linen and cotton threads among expensive silk threads were woven under the name of "Helali." In the 1950s, a new version of "Buldan twisted" namely as "Hoşgör" was woven by silk-cotton yarn blend. After 1970, this fabric is known as "Şile fabric." However, after 1970, when this species began to be recognized as "Şile cloth." the name of the district was added to the product and started to be called "Buldan Bezi" or "Buldan Bend" [55]. At Present, these plain weaves designed as lined, bordered, and square are embroidered with a variety of yarns and are utilized in clothing designs.

Buldan twisted cloth was used as underwear, especially as flannels. Nowadays, it is mostly used for the outer garments where the viscose fibers and shiny glitter materials are utilized. This product, also known as "Bürümcük." is produced with a plain weave connection (Fig. 4). Highly twisted weft yarns are preferred in this product. However, it is possible to come across products where high twist is used in both weft and warp. Hot soapy washing provides this fabric with wrinkling feature [55, 56]. In Buldan twisted clothes, thermal insulation is provided by the air layer between the human body and the Buldan fabric, which shrinks according to the twist

Fig. 4 Collarless Crepe
"Bürümcük" Shirt (This
textile product still being put
in Dowry Chests in Buldan)

rate of the yarns after washing at boiling temperature. It is also known that Buldan
weaves are very elastic because of twisted yarn hence covering the body with their
elasticity. The air inside the wrinkles of the 100% cotton fabric provides cool feeling,
high sweat absorption, and perspiration retardance [56]. Today the clothes made of
Buldan twisted fabrics are attracted by textile consumers due to their sustainable
structure, cooling, comfortable, healthy, and natural features. According to the local
people of the town, Collarless Shirt made of Traditional "Bürümcük" or in another
locally named "Hoşgör Fabric" is still a part of the bride and groom's dowry. The
other traditional Buldan products are also mentioned in the following part.

4.2 Other Traditional Buldan Products

Peshtemals (type of loin cloth or waist cloth and can be spelled as pesh-tamal, pestamal, or pestemal and spelled as peştemal in turkish language) are one of the important traditional products of Buldan woven fabrics. These clothes are called as "top (üstlük in Turkish language)" (for top half of the body) and "half (yarım in Turkish language)" (for the lower half of the body) (Figs. 5 and 6). Both textile products made from this traditional Buldan fabric are designed to cover most of the body. The first one is used to cover the upper part of the human body as shown in Fig. 6, and the second product is used to cover the lower part of the body (Fig. 6). The utilization of peshtemal as a traditional garment is not observed frequently in the town however elderly women living in Buldan frequently prefer them for covering their skirts, head, and body [56]. The economic situation of the person could be understood from the raw material of those loincloths. Wealthy families used to prefer those made of 100% silk, while the workers could afford to buy the ones made of 100% cotton. Today women living in rural areas of the district are still using them for covering their heads and hips (Fig. 6). However, due to the high cost of sourcing silk fibers, recently such products are often made from 100% viscose or cotton–viscose fiber blends.

Fig. 5 Traditional Buldan Clothes for top half of the body called as "top (üstlük)" and the lower half of the body called as "half (yarım)"

Fig. 6 A local woman in Buldan wearing the Traditional Buldan Clothes for "top (üstlük in Turkish language)" (for top half of the body) and "half (yarım in Turkish language)" (for the lower half of the body)

In the study carried out by Atalayer, it was declared that loincloths woven in Buldan-Denizli in Turkey can be divided into three section, namely, bath, tradesmen, and clothing loincloths [57]. And these loincloths may be striped silk or cotton fabrics [57]. Önlü stated in her study that women used to tie their loincloths around their waists over their clothes when they went out [58]. It was also mentioned in the early literature that bath loincloths are classified into men and women loincloths. Man generally use them around their naked waist, while women wear them starting from their armpits. This is the reason of women loincloths being wider and bigger than men loincloths [59]. Atalayer declared that technical properties of a loincloth woven in Buldan as 82 cm width, 120 cm length with the warp density of 13.2 and the weft density of 16 per cm. It was also stated in the study that fiber types of warp and weft are cotton, weaving type is plain weave [60].

Peshtemals have also been used as the irreplaceable element of Hammam Culture and Turkish baths since Ottoman times. New sophisticated weaving techniques were applied with new materials like cotton and silk especially in the seventeenth century. Buldan had the advantage of taking strengthening as an important textile center [56]. The handwoven peshtemals are still woven with handlooms as "original pieces" in Buldan. Figures 7 displays a hand made peshtemal with the original fringes (tufts, tassels) produced in Buldan. Figure 8 exhibits a traditional Buldan peshtemal robe example (also can be used as bathrobe) and a Buldan baby robe example. Different patterns of woven clothes may be provided with the most preferred colors of blue, green, yellow, orange, beige, and purple [61, 62].

Additional to towels and peshtemals, some more traditional products such as handkerchiefs, fine shirting fabrics, white linings and silk fabrics, home textile products such as bed sheets, tablecloth, pillow cover, and cushion covers are also produced in Buldan. Faroqhie investigated the textile products of Ottoman Empire in his study where bed sheets, table clothes, sheets, and various cotton-silk handkerchiefs and cotton curtains were mentioned to be produced in Buldan at the end of nineteenth century [63]. Uğurlu *et. Al.* mentioned about the "Krapons" with crepe weaving which are known to be used in school uniforms in Buldan district [64]. On the other hand, cotton or silk striped fabrics with high twisted yarns locally named "Hoşgör

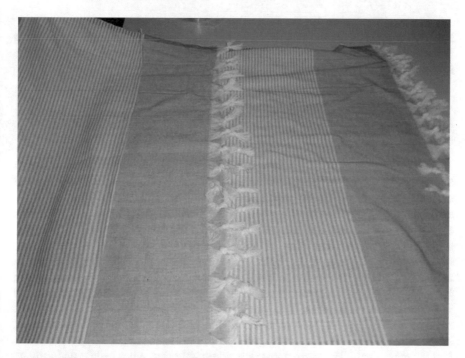

Fig. 7 Traditional Buldan peshtemal with fringes (tufts, tassels)

(a)   (b)

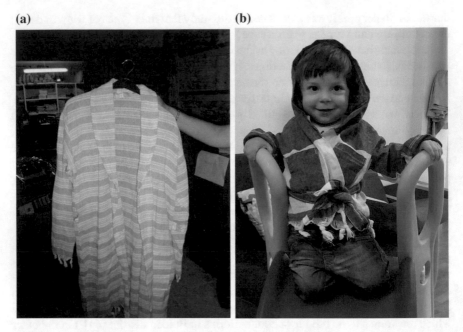

Fig. 8 A traditional Buldan peshtemal robe (also can be used as bathrobe, on the left) and a Buldan baby robe (on the right)

Fabrics" were also declared as popular products among the local people. Adding fringes and chain embroidery to the natural Buldan fabrics were alternatives for the local textile products [64].

5 Weaving Machines Utilized in Buldan

The first weavings in Buldan were woven on primitive hand looms found in almost every home. Production has been attempted by using Pit Handlooms firstly. In these systems, there are pedals attached to the frames that carry forces in the pit where the weaver sits, and behind the weaver there is a tree with a hole in the middle for stretching and loosening the warp. To prevent the woven fabric from shrinking, a tool called "metit." consisting of two pieces of laths with nails, is used instead of tweezers. It is known that people working in pit looms are frequently caught in rheumatic diseases. Therefore, it has been observed that hand looms working on a flat surface were preferred over time. Hence, the manual hand weave looms started to be utilized in the town. A hand loom is a simple machine used for weaving. The main parts of the manual hand weave loom may be listed as warp beam, cloth roller, heald shaft, healds, reed, slay, temple mill, cop, shuttle, beating shaft, pattern board, and the main shaft. The warp yarns pass alternately through a heddle and through a space between the heddles (the shed). Handlooms in Buldan are specific with the shuttle movement where the shuttle is thrown through the shed with the aid of whip. There is a manual hand system and a pedal system in the machines. Thanks to its synchronized work, weaving can be achieved.

In the manual handlooms, a weft package called a cop is added on the spindle in the shuttle and yarn is drawn from the eye of the shuttle [66]. Of course, there are some limitations of hand weave looms when compared with the automatic looms. For example, when the supply package is finished, the weaver takes out the empty package and puts a new one in the shuttle, draws the yarn from the shuttle eye and re-starts the loom again. This may be a serious time waste for the weaver [65, 66]. The basic principle of weaving for hand weave loom is the same with the semi- and fully automatic weaving machines. Sequence of operations in weaving (handloom) may be sorted as sizing, warping, drawing-in, denting, and weaving processes, respectively. Take-up and let-off motions are the secondary motions of the handloom. Figure 9 exhibits passing the warp yarns through the heddles in process performed by a craftsman weaver (Mr. Hasan KÖMÜRCÜOĞLU) in Buldan.

The warp threads are separated into two sheets (layers) by lifting some of the heald shafts up and by lowering others [66]. Because each warp yarn from the beams passes through an eye of the heald shafts; the related warp ends are up and down where an opening emerges. This is the time of shedding and there comes the next primary mechanism called "picking." A shuttle carrying the weft yarn is inserted through the shed by picking mechanism. As the shuttle reaches the other side of the loom, shed opening begins to close. Once the weft is inserted through the shed, beat-up motion occurs. "Reed" pushes the inserted weft into the edge of the already woven cloth which is named as "fell of the cloth." In here, the reed takes it motion from a crank shaft.

45 handlooms were identified in a study carried out in 2009 at Buldan [67]. 31 of these handlooms were in operation, and the rest of them were non-operational though preserved by their owners [67, 68]. Currently, 3000 motorized weaving looms, 30 hand weaving looms and 1250 embroidery machines are operating or ready to work in Buldan district center. According to the statements of the people of the district, there are around 20 hand weaving looms that are actively used today in Buldan. The rest of the handlooms are being kept by the owners although they are not actively used. Many different textile products produced by these actively operated hand weaving looms decorate people and their homes. After having explained the handloom weaving machine and its working principle in this part, information about "embroidery in Buldan Clothes" is given in the next part.

6 Embroidery in Buldan Clothes

Embroidery technique was applied to ancient wears and garments in humankind history. It may be created by utilizing needle or crochet on leather cloth or any other material such as felt produced by a variety of threads or fibers. Decorative elements

Fig. 9 The craftsman weaver (Mr. Hasan KÖMÜRCÜOĞLU) passing the warp through the heddles (heald eye)

have always been used for the aim of embroidery in Turkish culture. However, their images have variated throughout the history due to geographical and natural changes as well as due to interactions with other cultures and religions [69]. The embroidery of Turkish Republic period inherits its existence from Ottoman Empire. The gifts given and taken in diplomatic relations in Ottoman palace, the goods lefts during conquests, transaction coming from trade relations, Istanbul becoming an antic bazaar in nineteenth century influenced the development of embroidery. Embroidery was not only used for decoration purpose in Ottoman Empire but also reveals the social status of who wears it. It is mentioned in the literature that a noble woman is differed from the one who was not, a married one from a single, a person coming from İstanbul from the one coming from the countryside [70, 71].

Due to the changes in social order and lifestyle, there have become some changes up to today's embroidery designs. Semi-automatic machines and full-automatic machines in the 1980s resulted with the emerge of industrial concept also in embroidery. Today fully automatic machines benefitted from computerized embroidery programs are used in embroidery which reflects the creativity of Turkish people. Buldan weaves may be decorated with local motifs such as Buldan lake, bird's eye, black almond top. Some geometrical patterns, stripes are also used. Suzani (framework) a hand embroidery technique is a well known Buldan work which can also be applied to machines. The threads for embroidery may be from natural or from synthetic fibers. Although silk fiber threads were used more frequently in the past, viscose fiber thread (also called as floss silk thread or floss thread in the Buldan district) and synthetic polyester thread (also called as orlon thread in the Buldan district) have been used as embroidery threads recently due to the decrease in silk production in the town and the high cost of silk fiber. Today there are many local workers who owns professionality in using these special Suzani machines in the district factories [72]. Figures 10, 11, 12, 13, 14, and 15 reveal the examples of Buldan fabrics with the traditional embroideries. Next part includes an overview about some previous studies in the literature about Buldan cloth.

Fig. 10 Embroidery process with Suzene Hoop Machine on Buldan Cloth

Fig. 11 A special bed cover named "Sıvama" embroidered with "Intense Suzene Embroidery" on Buldan Fabric

7 Some Previous Studies in the Literature About Buldan Cloth

Some of the available resources related to history of Buldan cloth, fabric structure, applications of its modern usage in fashion are mentioned below. For instance, Aytaç and Hidayetoğlu conducted a study investigating the Buldan fabrics produced by manual looms [73]. Yakinol et al. conducted an analysis about the weaving looms, working people and environmental conditions of traditional Buldan hand weaving in an ergonomic aspect [74]. Status of innovation activities of 59 textile companies in Buldan was determined in the study of Ertuğrul and Utkun. The researchers conducted a survey study [75]. A survey study was conducted among all small workshops, small business centers and factories in Buldan which was based on the previous

(a)   (b)

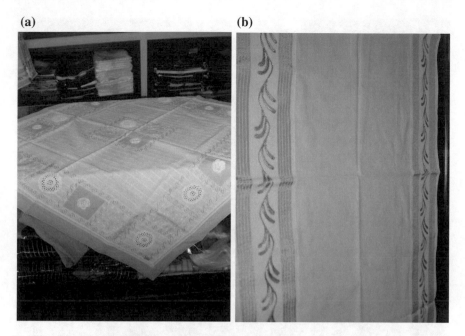

Fig. 12 Embroidered tablecloths made from Buldan fabric

Fig. 13 Embroidered coffee tablecloth (chat tablecloth) made from traditional Buldan fabric

Fig. 14 Embroidered tablecloth set (tablecloth and table napkins) made from traditional Buldan fabric

work of Ertuğrul and Utkun's inventory study in 2009 with the support of Buldan chamber of commerce in 2015. The survey questions were mostly about the working conditions, the yarn types utilized by the local business centers, distribution of loom types by the regions, raw material sources, innovation and exporting activities of enterprises [67, 68]. Utkun and Kırtay conducted another research about the steps of hand weaving process and the main factors influencing the quality and the price of hand weaving [76]. Özdemir and Utkun compared the profit margin of hand-woven products and those woven with automatic looms in Buldan textile. In here, the authors aimed to point out the importance of production in hand looms. The unit costs of "scarf" [made from silk, viscose (also called as floss silk or floss in the Buldan district), cotton and linen (flax)] woven in hand looms and woven with automatic looms were calculated, and it was concluded that profit margin of hand-woven scarf was much higher than those woven in automatic looms [77]. Historical process of Buldan weaving was revealed and the textile processes' innovation was also mentioned in Yılmaz's study [78]. In a study carried out by Kahvecioğlu, some mechanical properites of Buldan fabrics were investigated. The researcher compared the fabric tenacity and tearing strength properties of Buldan fabrics woven on semi-automatic looms and the fabrics woven on full-automatic looms. It was observed the fabric groups woven in Buldan revealed lower fabric tensile properties however they indicated higher fabric tearing strength in warp and weft wise [79]. Öztürk

Fig. 15 Embroidered nesting tablecloth and tablecloth made from traditional Buldan fabric

and Çoruh made a study related to utilization of Buldan clothes in fashion industry. Modern clothes were designed with Buldan fabrics and adorned in various ways [80].

Gündoğan et al. conducted a study where they tried to obtain different shades and colors with combination of heather and madder dye [81]. The simultaneous mordanting method was used for natural dyeing. The authors concluded that different color shades with sufficient fastness properties may be provided on Buldan fabric using different mordants with the natural dyes and heather [81]. Akduman performed a study where 100% Buldan fabrics were naturally dyed with madder and wall nut shells with different mordant. CIE $L*a*b*$ values and color differences of dyed samples with different mordants were evaluated in terms of light fastness and other fastness values [82].

Since the Buldan weavings are known to be special for providing comfortable clothes, Akgün et al. (2020) conducted a study about the effects of different fabric structural parameters of Buldan fabrics made of 100% cotton and made of cotton-Tencel fiber blend on comfort characteristics such as thermal resistance, thermal absorptivity, water vapor permeability and air permeability. Test results revealed that 100% cotton Buldan fabric properties had similar properties with those made of cotton/Tencel Buldan fabrics. Additionally, 100% cotton Buldan fabrics provided the lowest thermal absorptivity which means they give warm feeling among the others [83]. Our next part gives some examples of local workshop centers for hand weaving in Buldan.

8 The Examples of Local Workshop Centers for Hand weaving in Buldan

Although the industrialization in textile sector has contributed negatively on the handwoven items' production amount in Buldan, there are still some workshops keeping traditional production alive where the handwoven products are being sold as personal needs or as a souvenir. For example, traditional scarves, ties, shawls, peshtemals are woven with a hand loom in the entrance ground floor of "Evliyazadeler Konağı" (Evliyazadeler mansion in Buldan). For instance, while you are drinking your traditional Turkish coffee on the upper floor of this mansion, your handwoven product is made ready to be purchased on the lower floor of the mansion. Traditional products made of silk, viscose (also called as floss silk or floss in Buldan), cotton, and linen are also nice options as a local souvenir.

"BELSAM" (Buldan Handicraft Center) is another handicraft center where special woven products are made with cotton, silk [obtained from bombyx mori (Domestic silk moth)], viscose (also called as floss silk or floss in Buldan), and bamboo fiber threads with the utilization of hand weaving looms. This handwoven center can also be visited by the tourists where there are hand weaving looms controlled by old weaving masters. In here, the products produced by a weaving master (Mr. Yaşar Gürbıyık), who is now 70 years old, attract the attention of tourist groups. Here, production is carried out by trying to adhere to the traditional weaving characteristics of the region. In the center where narrow width weavings are produced intensively. Other products such as shawl, peshtemal, crepe weaving, dowry shirt, especially covered 'Half' veils while going to weddings and "loincloths" tied to the bottom are important products.

Silk hand weaving is very limited today in Buldan. In BELSAM, narrow width woven fabric can be produced from silk threads on most of the hand looms (Fig. 16).

Fig. 16 Production of silk fiber fabric using handloom in BELSAM

Silk products have been of particular importance for the Buldan people since ancient times. It has been stated in the literature that especially the people of Buldan attach importance to silk products in terms of being ostentatious at weddings [64]. Figures 17, 18, and 19 indicate some Buldan fabric products woven at the handlooms in BELSAM.

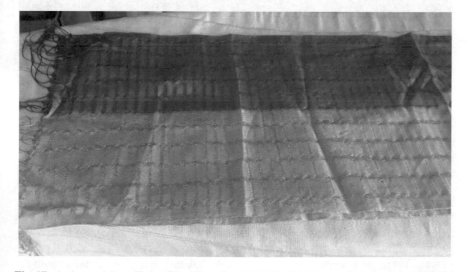

Fig. 17 A viscose (also called as floss silk or floss in Buldan) Buldan scarf produced in BELSAM

Fig. 18 Cotton woven peshtemal samples produced in BELSAM

Fig. 19 A sample of cotton woven peshtemal with high twisted yarns in BELSAM

To sum up, today hand weaving and crafting sectors are still one of the most important sources of income in many different countries such as India, Ethiopia, Nigeria, Sri Lanka, Bangladesh, Egypt, Thailand, etc. [84–88]. The craft designs on the materials reflect oriented aesthetics as well as their local identity. Hand weaving and its handwoven products have an important place in the lives of Anatolian people as well. There are many regional areas beside Buldan town where hand weaving is carried out within Turkey. Kızılcabölük and Babadağ towns in Denizli city, Ödemiş and Tire towns in İzmir city, Yesilyurt and Üzümlü towns in Muğla city, Aydın, Uşak, Afyon and Kütahya cities are some of these examples. The important thing is to contribute to sustainable textile production by protecting hand weaving and promoting handwoven products in these towns and cities where hand weaving has been actively used since the past.

9 Conclusion

Sustainability includes the industrial development without harming the environment and consuming natural resources as well as creating new strategies for future generations. Sustainability should be considered in three aspects as economical, environmental, and social sustainability. Due to the significant increase in the textile consumption, consumption of energy sources and using chemicals which harm the environment in the industrial processes resulted with the requirement of sustainable solutions in textile sector as well. Weaving with the handlooms may be a good alternative for the enhancement of sustainable textile products as well as creating a social sustainability where the weaving craft is passed down and taught from generation to generation. Hand weaving and handwoven products have an indispensable place in the lives of Anatolian people. Craft designs on the materials incorporate different aesthetic elements as well as local identities. Therefore, hand weaving has been carried out for centuries in the territory of Turkey. There are many regional areas where hand weaving is still carried out with in Turkey. Buldan, Kızılcabölük and Babadağ towns in Denizli city, Ödemiş and Tire towns in İzmir city, Yesilyurt and Üzümlü towns in Muğla city, Aydın, Uşak, Afyon, and Kütahya cities are some of these examples. Buldan fabrics produced in Buldan town of Denizli city in Turkey have been produced for centuries and have become famous as Buldan fabric.

Buldan cloth is not just a way of production but it is the living tradition which reflects the district's culture and history. Special handwoven products are still designed by the specialised craftsmen in the town. Some of the woven products can also be produced by automatic weaving machines in Buldan. In either way, there is always labor requirement for the final product. This is very important for the people of the district in terms of providing employment and job opportunities for many people. The rapid changes in the field of science and technology in the twenty-first century significantly affect the life of society. There have been positive developments within these rapid changes in Buldan as well. It is also observed that negative developments are also experienced in Buldan cloth, which are tried to be kept alive. Weaving, which is the most prominent and known traditional handicraft in Buldan, also resists technological advances. Over time, hand looms are mostly replaced by motorized looms. Again, silk weaving production has decreased compared to the past, due to reasons such as the decrease in silk production in the town and the high cost of silk. Some strategies can be realized for traditional Buldan cloth hand weaving to have the value it deserves. For Buldan weaving to be survived, Buldan cloth must complete the branding process. All kinds of innovations and developments that will increase the attractiveness of Buldan woven fabrics will enable these traditional products to be passed on to tomorrow's generations. Of course, this process requires knowing the history of these products very well and digesting the whole process. In addition, it is necessary to get the opinions, expectations, and suggestions of the end consumers about Buldan fabric products and to take steps to meet these expectations and make good touches and improvements to the products. Traditional Buldan products such as "Hoşgör" should be identified, and opportunities should be sought to increase the volume of these products in trade. More cooperation and collaborative works should be done between the different departments of universities such as textile engineering, fashion and design departments and the enterprises in Buldan. Associations, Unions, and chambers should break the perception of contract manufacturing and support the creation of special brands for manufacturers. The number and quality of hand looms should be increased for added value of Buldan cloth/woven apparel products. It will be considerable to open a textile museum in Buldan at the earliest opportunity. There may also be a library where resources about Buldan textile will be exhibited. It can also advised to provide trainings and supports to the producers in Buldan for combining traditional hand weaving with innovated products. Within the scope of the steps that can be taken towards branding in Buldan, efforts should be made to establish quality standards in Buldan Cloth and to expand its international recognition. Here, it is of great importance to work customer-oriented, especially in fabrication production. Opening stores and direct marketing in various countries can help these sustainable products find more places in world markets. In the meantime, it is necessary to benefit from all incentives and supports related to export and marketing.

The use and recognition of handwoven Buldan fabrics will increase with the positive contributions of all the different approaches mentioned here and other approaches that will work. In this way, while benefiting both the economy of the region and the country's economy, great added value will be provided in producing sustainable textile products. So, this also ensures and adds up to the sustainability efforts that traditions, crafts, and experience are passed down from generation to generation, while also providing employment for people who do this work. A beautiful and sustainable future will only come to life by choosing and using more sustainable products.

Acknowledgements The authors would like to thank to especially to Halil Baştürkmen (President) and Ayhan Emirdağ from "Buldan Chamber of Commerce." BELSAM (Buldan Handicraft Center) and hand weaving masters Mr. Habip PEKÖZ and Mr. Hasan KÖMÜRCÜOĞLU for their great contributions.

References

1. Unal F, Avinc O, Yavas A, Eren HA, Eren S (2020) Contribution of UV technology to sustainable textile production and design. In: Sustainability in the textile and apparel industries. Springer, Cham, pp 163–187
2. Unal, F, Yavas A, Avinc O (2020) Sustainability in textile design with laser technology. In: Sustainability in the textile and apparel industries. Springer, Cham, pp 263–287
3. Eren HA, Yiğit İ, Eren S, Avinc O (2020) Sustainable textile processing with zero water utilization using super critical carbon dioxide technology. In: Sustainability in the textile and apparel industries. Springer, Cham, pp 179–196
4. Yıldırım, FF, Yavas A, Avinc O (2020) Printing with sustainable natural dyes and pigments. In: Sustainability in the textile and apparel industries. Springer, Cham, pp 1–35
5. Eren HA, Yiğit İ, Eren S, Avinc O (2020) Ozone: an alternative oxidant for textile applications. In: Sustainability in the textile and apparel industries. Springer, Cham, pp 81–98
6. Yıldırım FF, Avinc O, Yavas A, Sevgisunar G (2020) Sustainable antifungal and antibacterial textiles using natural resources. In: Sustainability in the textile and apparel industries. Springer, Cham, pp 111–179
7. Eren S, Avinc O, Saka Z, Eren HA (2018) Waterless bleaching of knitted cotton fabric using supercritical carbon dioxide fluid technology. Cellulose 25(10):6247–6267
8. Setthayanond J, Sodsangchan C, Suwanruji P, Tooptompong P, Avinc O (2017) Influence of MCT-β-cyclodextrin treatment on strength, reactive dyeing and third-hand cigarette smoke odor release properties of cotton fabric. Cellulose 24(11):5233–5250
9. Yavaş A, Avinc O, Gedik G (2017) Ultrasound and microwave aided natural dyeing of nettle biofibre (Urtica dioica L.) with madder (Rubia tinctorum L.). In: Fibres and textiles in Eastern Europe
10. Kurban M, Yavas A, Avinc O, Eren HA (2016) Nettle biofibre bleaching with ozonation. De Redactie 45
11. Avinc O, Erişmiş B, Eren HA, Eren S (2016) Treatment of cotton with a laccase enzyme and ultrasound. De Redactie 55
12. Eren HA, Avinc O, Erişmiş B, Eren S (2014) Ultrasound-assisted ozone bleaching of cotton. Cellulose 21(6):4643–4658
13. Davulcu A, Eren HA, Avinc O, Erişmiş B (2014) Ultrasound assisted biobleaching of cotton. Cellulose 21(4):2973–2981
14. Gedik G, Avinc O, Yavas A, Khoddami A (2014) A novel eco-friendly colorant and dyeing method for poly (ethylene terephthalate) substrate. Fibers Polymers 15(2):261–272
15. Gedik G, Yavas A, Avinc O, Simsek Ö (2013) Cationized natural dyeing of cotton fabrics with corn poppy (papaver rhoeas) and investigation of antibacterial activity. Asian J Chem 25(15)
16. Avinc O, Eren HA, Uysal P (2012) Ozone applications for after-clearing of disperse-dyed poly (lactic acid) fibres. Colorat Technol 128(6):479–487
17. Avinc O, Eren HA, Uysal P, Wilding M (2012) The effects of ozone treatment on soybean fibers. Ozone: Sci Eng 34(3):143–150
18. Karakan Günaydin G, Palamutcu S, Soydan AS, Yavas A, Avinc O, Demirtaş M (2020) Evaluation of fiber, yarn, and woven fabric properties of naturally colored and white Turkish organic cotton. J Textile Inst 111(10):1436–1453
19. Yıldırım FF, Yavas A, Avinc O (2020) Bacteria working to create sustainable textile materials and textile colorants leading to sustainable textile design. In: Sustainability in the textile and apparel industries. Springer, Cham, pp 109–126
20. Unal F, Yavas A, Avinc O (2020) Contributions to sustainable textile design with natural raffia palm fibers. In: Sustainability in the textile and apparel industries. Springer, Cham, pp. 67–86
21. Unal F, Avinc O, Yavas A (2020) Sustainable textile designs made from renewable biodegradable sustainable natural abaca fibers. In: Sustainability in the textile and apparel industries. Springer, Cham, pp 1–30

22. Kumartasli S, Avinc O (2020) Important step in sustainability: polyethylene terephthalate recycling and the recent developments. In: Sustainability in the textile and apparel industries. Springer, Cham, pp 1–19
23. Kumartasli S, Avinc O (2020) Recycling of marine litter and ocean plastics: a vital sustainable solution for increasing ecology and health problem. In: Sustainability in the textile and apparel industries. Springer, Cham, pp 117–137
24. Fattahi FS, Khoddami A, Avinc O (2020) Sustainable, renewable, and biodegradable poly (lactic acid) fibers and their latest developments in the last decade. In: Sustainability in the textile and apparel industries. Springer, Cham, pp 173–194
25. Gedik G, Avinc O (2020) Hemp fiber as a sustainable raw material source for textile industry: can we use its potential for more eco-friendly production? In: Sustainability in the textile and apparel industries. Springer, Cham, pp 87–109
26. Kalayci E, Avinc O, Yavas A, Coskun S (2019) Responsible textile design and manufacturing: environmentally conscious material selection. In: Responsible manufacturing: issues pertaining to sustainability (2019), 1
27. Günaydin GK, Yavas A, Avinc O, Soydan AS, Palamutcu S, Şimşek MK, Dündar H, Demirtaş M, Özkan N, Niyazi Kıvılcım M (2019) Organic cotton and cotton fiber production in turkey, recent developments. In: Organic cotton. Springer, Singapore, pp 101–125
28. Soydan AS, Yavas A, Günaydin GK, Palamutcu S, Avinc O, Niyazi Kıvılcım M, Demirtaş M (2019) Colorimetric and hydrophilicity properties of white and naturally colored organic cotton fibers before and after pretreatment processes. In: Organic cotton. Springer, Singapore, pp. 1–23
29. Günaydin GK, Avinc O, Palamutcu S, Yavas A, Soydan AS (2019) Naturally colored organic cotton and naturally colored cotton fiber production. In: Organic cotton. Springer, Singapore, pp. 81–99
30. Palamutcu S, Soydan AS, Avinc O, Günaydin GK, Yavas A, Niyazi Kıvılcım M, Demirtaş M (2019) Physical properties of different turkish organic cotton fiber types depending on the cultivation area. In: Organic cotton. Springer, Singapore, pp. 25–39
31. Kalaycı E, Avinç OO, Bozkurt A, Yavaş A (2016) Tarımsal atıklardan elde edilen sürdürülebilir tekstil lifleri: Ananas yaprağı lifleri." Sakarya Univ J Sci 20:2, 203–221
32. Avinc O, Khoddami A (2009) Overview of poly (lactic acid) (PLA) fibre. Fibre Chem 41(6):391–401
33. Avinc O, Khoddami A (2010) Overview of poly (lactic acid) (PLA) fibre. Fibre Chem 42(1):68–78
34. Palamutcu S (2017) Sustainable textile technologies. In: Muthu S (eds) Textiles and clothing sustainability. Textile science and clothing technology. Springer, Singapore
35. Annapoorani SG (2017) Social sustainability in the textile industry. In: Muthu S (ed) Sustainability in the textile industry. Springer, Singapore, pp 57–78
36. Muthu SS (2017) Evaluation of sustainability in textile industry. In: Muthu S (ed) Sustainability in the textile industry. Springer, Singapore, pp 9–15
37. Sherborne A (2009) Sustainability through the supply chain. In: Blackburn RS (ed) Sustainable textiles life cycle and environmental impact. Woodhead Publishing Limited, CRC Press, U.K, pp 1–32
38. Mulder K, Ferrer D, Lente H (2011) What is sustainable technology? Perceptions, paradoxes and possibilities. England: Greenleaf Publishing Limited. 978-1-906093-50-1. http://www.greenleaf-publishing.com/technology
39. Devrent N, Palamutçu S (2017) Organic cotton. Paper presented at the International Conference on Agriculture, Forest Food Sciences and Technologies, Cappadocia / Turkey, 15–17 May 2017.
40. Choudry AKR (2014) Environmental impacts of the textile industry and its assessment through life cycle assessment. In: Muthu S (ed) Roadmap to sustainable textiles and clothing. Springer, Singapore, pp 1–39

41. Islam S, Mohammad F (2014) Emerging green technologies and environment friendly products for sustainable textiles. In: Muthu S (ed) Roadmap to sustainable textiles and clothing. Springer, Singapore, pp 63–82
42. Niinimäki K (2015) Ethical foundations in sustainable fashion. Textiles Cloth Sustain 1(1):3
43. Uludağ İhracatçı Birlikleri Genel Sekreterliği Ar-Ge Şubesi, Ev Tekstili Sektörü Raporu, https://uib.org.tr/tr/elektronik-kutuphane.html, Accessed: 03 Nov 2018
44. Saxena S, Raja ASM (2014) Natural dyes: sources, chemistry, application and sustainability issues. In: Muthu S (ed) Roadmap to sustainable textiles and clothing. Springer, Singapore, pp 37–80
45. İşmal ÖE (2017) Greener natural dyeing pathway using a by-product of olive oil; prina and biomordants. Fibers Polymers 18(4):773–785
46. Khattab TA, Abdelrahman MS, Rehan M (2020) Textile dyeing industry: environmental impacts and remediation. Environmental Science and Pollution Research, pp 1–16
47. Altinöz H, E. (2012) Denizli'de Öncü Bir Sanayi Kuruluşu: Göveçlik İplik Sanayi ve Anonim Şirketi (1972–2008). Belgi Dergisi 4:390–409
48. Travel notes. https://www.storiesandobjects.com/blogs/travel-notes/buldan-turkey, Accessed 2 April 2018
49. Atik N, Erdem ZK (2002) TÜBA-TÜKSEK Buldan (Denizli) Arkeolojik Belgeleme Çalışması (Archaeological Documentation Work), Türkiye Bilimler Akademisi - Türkiye Kültür Envanteri (TÜBA-TÜKSEK) Pilot Bölge Çalışmaları, 1/2, 1–7, TÜBA-T.C. Kültür Bakanlığı Yayını, İstanbul
50. Özer ML (2016) Yöresel Bir Dokuma: Buldan Bezi. In: Paper presented at the international conference, Bezce 7. Uluslararası Tekstil Konferansı "Anadolu'ya Dokunan Bezler, İstanbul/21–23 March 2016, pp 199–202
51. Valiliği D (2020) (Governorship of Denizli), Offical web site of Denizli Valiliği İl Kültür ve Turizm Müdürlüğü. http://www.denizlikulturturizm.gov.tr/tr/content.asp?id=371 [26.10.2008/Date of access: 14 08 2020
52. Ayata S (1988) Kasabada Zenaat Üretimi ve Toplumsal Tabakalaşma (Buldan) (Craft Production and Social Stratification in the Buldan Town. ODTÜ Gelişme Dergisi 15(1–2)
53. Denizli Ticaret Odası (Denizli Chamber of Industry), Official web site of Denizli Chamber of Industry. https://www.dto.org.tr/. Date of access: 20.10.2020
54. Texile in Life; Buldan weaves. https://texinlife.com/traditional-anatolian-buldan-weaves/. Date of access: 14.09.2020
55. Erdoğan ZTD, Arlı MTD (1996) Buldan dokumacılığı ve ilçede üretilen düz dokumaların bazı özellikleri üzerinde bir araştırma, A research on Buldan weaving and some properties of flat weaves produced in the district, Ankara University Institute of Natural Applied Science, Department of Home Economics, Ankara
56. The traditional craftsmanship of Buldan. https://kipsakes.com/blogs/heritage/the-traditional-craftsmanship-of-buldan. Accessed 10 July 2017
57. Atalayer G (1987) Bükülü Bez (Bürümcük) (Twisted Buldan Cloth), Türkiyemiz Dergisi; Journal of Türkiyemiz, 53:29–33
58. Önlü N (2010) Ege Bölgesi El Dokuma Kaynakları. Sanat Dergisi 17:47–60
59. Cillov H (1949) Denizli El Dokumacılığı Sanayi. İstanbul Üniversitesi İktisat Fakültesi, İktisat ve Neşriyat Enstitüsü No: 10. İsmail Akgün Matbaası, İstanbul
60. Atalayer G (1993) Dünden Bugüne Anadolu'da Kumaş Dokuma Sanatı, Türk Kültüründe Sanat ve Mimari, 64, İstanbul
61. Kalkanci M (2018) Different approaches in bathrobe manufacturing: new concept Pestemal bathrobes, an irreplaceable element of Hammam culture. Fibres Textiles Eastern Europe 26(4):130
62. Erdogan Z, Söylemezoglu F, Kahvecioglu H (2009) The use of traditional Buldan weavings in interior design. In: Paper presented at 4th international scientific conference proceedings of the 4th international scientific conference, Jelgava, Letonya

63. Faroqhi S (2004) Osmanlı dünyasında üretmek, pazarlamak, yaşamak. Yapı Kredi Yayınları 1904 Tarih-21 Toplumsal Tarih Araştırmaları Dizisi: 10. Acar Matbaacılık, İstanbul
64. Uğurlu A, Uğurlu SS (2006) Yörenin kültürel kimliği olarak Buldan Bezi. In: Paper presented at the international conference of Buldan symposium, 23–24 November, Buldan, Turkey, pp 275–280
65. Büken NRO (2005) El Dokumaciliğinin ve El Dokuma Tezgahinin Tarihçesi. El Dokuma Tezgahi Çeşitleri. Sanat Dergisi 8:63–84
66. Gandhi KHL (2019) The fundementals of weaving technology. In: Gandhi KHL (ed) In: Woven textiles: Principles, technologies and applications. Woodhead Publishing, Cambridge, England, pp 167–269
67. Ertuğrul İ, Utkun E (2009) Buldan Tekstil Sanayiinin Gelişimi ve Envanter Araştırması (Buldan Textile Industry Development and an Inventory Research) Ekin Basın Yayın Dağıtım-Bursa
68. Türköz N (2015) Buldan Tekstil Sektörünün Gelişimi ve Envanter Araştırması (Buldan Textile Industry Development and an Inventory Research) , Denizli, 99s
69. Göksel N, Kutlu N (2016) Decorative elements in Turkish garment culture from past to future: art of embroidery. J Textiles Eng/Tekstil Ve Mühendis, 23(103)
70. Akbil F (1992) Viyana Uygulamalı Sanatlar Müzesindeki Türk İşlemeleri ve Halılarından Örnekler. Kültür ve Sanat Dergisi 14:75–76
71. Sürür A (1976) Türk İşleme Sanatı, Ak Yayınları. Türk Süsleme Sanatları Serisi 4, 68, İstanbul
72. Bahar T, Baykasoğlu N (2017) Embroidery in Turkish culture. In: Recep E, Penkova R, Wendt JA, Saparov K, Berdenov JG (eds) Developments in social sciences. ST. Klıment Ohridski University Press, Sofia, Bulgaria, pp 203–214
73. Aytaç A, Hidayetoğlu MH (2004) Buldan Mekikli Dokumaları. Milli Folklor Üç Aylık Uluslararsı Halk Bilimi Dergisi 8(62):49–52
74. Yakınol ZE, İlleez AA, Güner M (2009) Buldan El Dokumacılığının Ergonomik Olarak İncelenmesi, Paper Presented at 1. Uluslararası 5. Ulusal Meslek Yüksekokulları Symposium, Konya, Turkiye
75. Ertuğrul İ, Utkun E (2011) A research about determining the innovation activities of textile companies in Denizli. International entrepreneurship congress, İzmir University of Economics, İzmir, Buldan, pp 5–12
76. Utkun E, Kırtay E (2009) El Dokumalarında Kaliteyi ve Fiyatı Etkileyen Faktörler, 1. Uluslararası 5. Ulusal Meslek Yüksekokulları Sempozyumu, Konya, Turkiye
77. Özdemir S, Utkun E (2014) Comparing the profit margin of textiles woven on a hand loom and automatic loom in the textile industry: case of Buldan city
78. Yılmaz S (2006) Tekstil Uygarlığı ve Buldan. A conference paper presented at Buldan Sempozyumu, Buldan, 23–24 November 2006, vol 1, pp 261–273
79. Kahvecioğlu H (2006) Buldan Dokumalarının Bazı Mekaniksel Özellikleri Üzerine Bir Araştırma, A conference paper presented at Buldan Sempozyumu, Buldan, 23–24 November 2006, vol 2, pp 623–634
80. Öztürk F, Çoruh E (2013) An application concerning utility of Buldan weaves in fashion industry, vol. 20, p 91
81. Gündoğan M, Avinç O, Yavaş A (2017) Dyeing of Buldan Fabric with the combination of common Heather (Calluna Vulgarıs) and Madder (Rubia Tinctorum, 2. In: International Mediterranean art symposium, 10–12 May 2017. Antalya/Turkey, pp 138–143
82. Sarı HK, Akduman Ç (2017) Dyeing of Buldan Fabrıcs with Walnut Shells (Junglans Regıa L.) and Madder (Rubia Tinctorum L.) 2. In: International mediterranean art symposium, 10–12 May 2017, Antalya/Turkey, pp 104–108
83. Akgun M, Gurarda A, Gunaydin GK, Çeven EK (2020) Investigation of the comfort properties of traditional woven fabrics with different structural parameters. Industria Textila 71(4):302–308
84. Temesgen AG, Turşucular ÖF, Eren R, Ulcay Y (2018) The art of hand weaving textiles and crafting on socio-cultural values in Ethiopian. Int. J. Adv. Multidiscip. Res 5(12):59–67
85. Henze M (2007) Studies of imported textiles in Ethiopia. J Ethiopian Stud 40 (1/2):65

86. Dissanayake DGK, Perera S, Wanniarachchi T (2017) Sustainable and ethical manufacturing: a case study from handloom industry. Textiles Clothing Sustain 3(1):2
87. Chantaramanee N, Taptagaporn S, Piriyaprasarth P (2015) The assessment of occupational ergonomic risks of handloom weaving in northern Thailand. In: Science and technology Asia, pp 29–37
88. Rani N, Bains A (2014) Consumer behaviour towards handloom products in the state of Punjab and Haryana. Int J Adv Res Manag Soc Sci 3(10):92–105

The Cultural Sustainability of the Textile Art Object

Marlena Pop

Abstract The various handloom techniques are the oldest artistic technologies of humanity. The typologies of archaic weaving techniques, as well as of the raw material, are very diverse on all the meridians of the world. Through the perpetuation of techniques, materials, archetypes and symbols, it has been possible to transmit to this day an entire material and immaterial heritage of fabrics, craft techniques, symbols and decorative motifs, for all the peoples of the world. All these form that patrimony of the archaic textile arts which can be described by the anthropological self-referentiality of the handloom. The material and intangible heritage must be preserved in museums, but capitalized creatively and intelligently so that future generations can understand the identity, history and great cultural diversity of the handloom. The modern concept of cultural sustainability of products is based on the principle of sustainability over time, with a small carbon footprint. The technical processes of making sustainable products in art and textile design require a slowdown in serialization, production and consumption, as well as an increase in added value and creative personalization, with emphasis on the visual semiotics of cultural identity. In this context, the creative design of textiles, oriented towards application areas such as fashion textiles and those for interior or architectural design, use the appropriate cultural tools for any field of sustainable creative industries. The works that arc the subject of this article proves that exploratory and experimental research in textile arts, conducted with students of the Department of Textile Arts and Textile Design at UAD Bucharest, was able to validate cultural tools and aesthetic material basis, allowing the creative to express, through specific visual language, an entire individual universe, an archetypal heritage identity, with sustainable cultural values.

Keywords Archaic handloom · Culture · Identity · Sustainability · Re-crafting · Arts

M. Pop (✉)
HDCCI, Humanities & Design Center for the Creative Industries, Bucharest, Romania

M. Á Gardetti and S. S. Muthu (eds.), *Handloom Sustainability and Culture*,
Sustainable Textiles: Production, Processing, Manufacturing & Chemistry,
https://doi.org/10.1007/978-981-16-5967-6_6

1 Introduction

Our mental stratifications, expressed through the collective imaginary, are true repositories of anthropological cultural archaeology that are transmitted from one generation to another through knowledge, education, culture, habits of civilization, customs and craftsmanship. The forms of expression of the imaginary are those that describe the meaning of the diachronic evolution of the ethnic diversity of identity artefacts. This level of knowledge of human identity has been reached through the anthropological self-referentiality because each culture is the repository of the past and/or of the knowledge of its own symbolic creativity.

The circular economy is the image of a global sustainability, but each person expresses themselves through their own cultural identity, to the sense of belonging to a nation and country, to their own history and spirituality. Why is there a need for a product culture and how is cultural identity expressed through the product? This is already becoming an issue for artists, designers and marketers, whose complex, interdisciplinary research will lead in the future to the development of the fourth pillar of the circular economy: the culture of sustainable products. In the context of these economic developments, the observation of creativity through the artistic productions of the world of visual arts and design in direct relation with their manufacturing technologies is discussed:

- *creativity in the context of material and immaterial cultural diversity and the existence of a tendency to highlight the patrimonial and archetypal cultural sustainability;*
- *the problem of identifying cultural creativity of the products in relation to the production technologies, the types of mediation of the reception relations of products with cultural sustainability;*
- *harmonization of cultural sustainability with the ecological material base of the products and with the logistic systems with traceability and transparency.*

These are some practical issues that directly target creativity in relation to the cultural sustainability of products, being an important topic of research in art and design in this post-industrial moment from an economic point of view and post-modernist from a cultural point of view. In any artistic field and at all times, art has its elements of self-referentiality. Throughout this research, we used the term "anthropological self-referentiality" to show that the entire cultural continuity of craft techniques, symbols and folk motifs passed down from generation to generation, defining an ethno-cultural area, can only be explained by the existence of this process of becoming aware of ethnic creativity.

The diacritical approach of research in the practice of textile decorative arts can only have historical, archaeological reference of weaving techniques, weaving tools, as well as archetypal cultural elements that are part of a visual semiotics of the creativity of the people of those places.

As Romanian archaeologists and historians say, "Neolithic cultures in the Romanian area are in line with the European direction in terms of technological choice,

Image 1 Pieces from the ceramics collection of the Vadastra Culture (5000–4500 BC) of the Museum of Romanian History, https://www.mnir.ro/, handloom reconstruction in Romanian Neolithic cultures within the UNA Bucharest project *"Time Maps. Real Communities—Virtual Worlds—he Experienced Past"*, 2011–2013

namely the vertical handloom with weights. The handloom weights of this period are pear-shaped or round and thick, each with a hole, shuttles were made of bone or wood and the raw materials used were hemp, flax and, towards the end of the period, wool".[1] In the reconstruction of the weaving techniques, a vertical handloom with weights was made (image 1), taking as model the 18 conical weights, archaeological pieces, exhibited at the Romanati Museum from Caracal, Olt county, Romania, as well as processing yarns of hemp, wool, dyed with tinctorial plants, specific to the area: celandine, buckthorn, St. John's wort, walnut shells, nettles and specific mordants: salt and cabbage juice.

Regarding the processing techniques of bast textile plants, flax and hemp, they were perpetuated and perfected, but were preserved in their archetype in Romanian villages, until the nineteenth century. Today's generations, not having a direct contact with the archaic processing of textile plants, are trying to reconstitute them in the Romanian villages from the historical regions of the country. The entire production of textile objects was in the care of women in all historical times. Thus, they had to take care of the whole process, from harvesting textile plants, to processing them, differentiated weaving of fabrics for work clothing and holiday costumes for the whole family, household fabrics (bed and table linen or for home decoration), the technical fabrics needed in the household or in agricultural activities. Throughout Romania, the ancestral processing of hemp followed this technology that became ritual and apotropaic because the different stages of processing and making fabrics

---

[1] Gheorghiu and Rusu [4].

were accompanied by incantations, stories, songs or games in group work. The archaic weaving involved the use of hemp yarn for both warp and weft, and the fabric was accompanied by true chants.[2]

2 Methodology/Team/Materials and Techniques

The approach of anthropological self-referentiality as a research topic in artistic practice aims to verify a continuity of artistic experience of the Romanian woman using the handloom, through explorations in the identity past of the Romanian archaic weaving, by knowing the references of the specialized literature, of the museums, in situ references from the old households, as well as by immersion experiments into a material, philosophical-symbolic imaginary and practically, through a laboratory in textile arts.

In any artistic field and at all times, art has its elements of self-referentiality. Observed from the Greco-Roman self-portrait, to the post-Renaissance mannerism or to the "art for art's sake" phenomenon, or Post-Impressionism (Vincent van Gogh— "I dream my paintings and then I paint my dreams") Symbolism, Surrealism (René Magritte is famous for his self-referentiality works. "The Treachery of Images"), to contemporary tautologies. In the twenty-first century, self-referentiality is present in all artistic media, or not, that pass through mass-media. Basically, the entire field of culture has become self-referentiality: literature is a metaliterature, visual art is self-referentiality of the personal event, advertising art is self-referentiality by definition, etc. Throughout this research we used the term "anthropological self-referentiality" to show that the entire cultural continuity of craft techniques, symbols and folk motifs passed down from generation to generation, defining an ethno-cultural area, can only be explained by the existence of this process of becoming aware of ethnic creativity. On the other hand, a collaborative research, such as that of a research team in which the team members are from the same field of expertise, cannot be appropriately developed if the method of concerted dialogue is not employed. It is necessary to develop dialogic skills in collaborative research because the communication skills of team members and the specific particularities of using visual and verbal languages can lead to obtaining intelligible artistic discourses in both modes of expression. During this research of self-referentiality in the textile arts in conjunction with the application of the design thinking method in the decorative arts, the dialogic development of communication was also aimed at, exactly in the sense presented above.

---

[2] The ethnographic reconstructions, made by teams of young ethnographers, even if they do not reflect the depth of textile plant processing techniques, they generate a practical understanding of the harshness of these techniques as well as the exhausting work of Romanian peasants who were in constant competition with the other women of the village so that every year, at Easter, at the church service, the whole family was dressed in new clothes, the house was decorated with new towels, linen sheets or wool rugs. https://www.youtube.com/watch?v=J8MKr_NfGHA&t= 662s https://www.youtube.com/watch?v=HAldsF6SCyE&t=742s.

No fragments of Neolithic or Eneolithic fabrics have been found in the Romanian archaeological area so far. But their non-existence does not prove the non-existence of weaving because in all Neolithic sites, from the Boian, Vadastra, Gumelnita, Petresti, Cris, Salcuta, Hamangia or Cucuteni cultures, weights specific to the vertical handloom were discovered and only the perishability of the textile material led to the non-existence of artefacts. In a Neolithic Petresti settlement, handloom weights were identified in each Neolithic household, the figurines of the Cucuteni culture are richly decorated, suggesting an advanced technique of weaving and sewing, and in the Vadastra culture, the ceramic pieces have symbols and textile decorative motifs that have been perpetuated until today in the creation of the traditional costume of Romanati. Also, in Vadastra culture were found ritual statuettes that are in themselves true samples of Romanati costume, even if they are over 5000 years old.

The sustainable artistic approach, using re-crafting through the design thinking method in the practice of textile decorative arts, and anthropological self-referentiality is the methodological key addressed in this project. Self-referentiality "presents the awareness by the author or artist of the creative process, of the tech-niques he uses".[3] Several authors argue that self-referentiality is a feature of post-modernism[4] delivered primarily by diminished creativity and developed by adver-tising and marketing systems, focused on profit and repeating messages appreciated by the public. As the French philosopher says, "Transcending self-referentiality is not only a question of changing one's own inquiry practice, it is also about ques-tioning a tendency to impose one's own regime of truth on others".[5] These are the negative aspects of self-referentiality. One of the positive aspects is that of being a vector of the continuity of cultural identity. Thus, through traditional household crafts, techniques, motifs, identity colour palettes, cultural symbols have been perpet-uated over the millennia. An entire visual semiotics is tributary to anthropological self-referentiality. Another method used in the project is "design thinking" for the decorative arts.

The "design thinking" method, generated as a phrase by Herbert A. Simon (1996), a renowned specialist in the sciences of knowledge, has been rediscovered and used as a design method by Stanford professors and popularized by Ideo in recent years, as a method of business development in design. Using this method, the design process is no longer linear, but turns into an iterative, spatial and empathic process, being centred in each stage on the requirements of the person who will use that product. Throughout the project called "VADASTRA - Self-referentiality for the re-crafting design", the research team used the first method, the analysis of anthropological self-referentiality for immersion in the memory of ethnic time and experiencing the artists' experiences in the environment of the theme, in order to know their own creativity related to the theme, and the "design thinking" method was used to develop iterations and concepts of modern re-crafting design products in the textile arts. The

---

[3] Collins dictionary, 2020, UK.

[4] Lawson and Noth [9].

[5] Foucault, 2000, https://www.lkouniv.ac.in/site/writereaddata/siteContent/202004021930365629 saroj_dhal_socio_FOUCOULT.pdf.

iterative process in design and decorative arts consists in: identifying the design problem, first of all by defining the needs for that product, choosing those methods and cultural working tools corroborated with technical and economic analysis tools that optimize the specificity of the product range, generating ideas and development of an ideatic prototype (idea sketch, idea modelling, CAD model, sketch folder). Virtual simulation of the product is required to see if it meets all needs.[6] All mistakes, inadequacies, problems that do not meet the needs are corrected and applied in a new design, along with what was innovative in the previous iteration.[7]

The research team consisting of a researcher, a professor, three 3rd year undergraduate students and a textile technician made an entire artistic and procedural approach, from the immersion in the material environment and spirituality of the theme, the Vadastra ethos, to the sustainable use of textiles and archaic hand weaving tools. The duration of the project was 3 months, from establishing the theme, to the physical and virtual layout of the textile art works.

The material and technique were hemp and vertical handloom. An attempt was made to reconstruct the Neolithic handloom, because the Vadastra culture is a Neolithic culture (5000–4500 BC) from the southern part of Romania, in the historical region of Oltenia, Romanati, and hemp with traceability of Romanian origin was used, both in warp and weft, in undyed yarn and some yarn dyed red and brown, Neolithic colours. In order to understand the whole process of re-crafting, the students watched films of ethnographic reconstruction of hemp processing and archaic weaving and made explorations in the imaginary and material environments of textiles.

Hemp processing is entirely sustainable because nothing is lost, everything is transformed for the benefit of man and nature. Hemp processing and weaving, the first female artistic technology, with the role of dressing, protecting and beautifying the whole family and home have their origins in the Neolithic. This archaic practice was maintained in Romanian villages until the middle of the twentieth century, and contemporary textile arts still preserve the dignity of beauty through monumental works to enhance family or public spaces.

3 Results and Discussions

The stages of the project and its results, intermediate and final, can be observed through iteration sketches, models and processing using graphic software, with intuitive interfaces, specific to VR simulations, presented below, as well as through the relevant images that accompany them.

1. Anthropological self-referentiality through emotional immersion in the reconstruction of the ethos of the Vadastra Culture (5000–4500), Olt Valley: visualization of specific artefacts, reconstruction of the vertical handloom, visualization

---

[6] Pop, 2018.
[7] Jin, 2006.

of the ethnographic reconstruction of archaic hemp processing and weaving (image 1). It was the stage of initiating research, the stage in which the students carried out a deep and diverse documentary research, from the knowledge related to Vadastra culture to the imaginary reconstruction of a semiotic route, the Vadastra Culture–Romanati Costume (image 2). Located on the Olt valley, on the way of the legions to Apulum, in the heart of Trajan's Dacia, the territory of Romanati county had a rich economic life both in antiquity and in the Middle Ages. "Here is Romula (Little Rome) where, "in the archeological researches carried out in the last decades in the Roman forts of Dacia (lower) Malvensis, some fragments of terra sigillata vessels were discovered".[8] [8] Name of the region, Romanati, with the Slavic suffix, "ati" derives from the name of a prince or county: Roman (ati)". Today the region is, due to climate change, in the process of desertification, having as a consequence and depopulation. Only the historical ethnographic and archeological value reminds of the rich Romanati region.

2.	The perception of personal emotions in the context of an informed creative imaginary and the expression of the first symbolic, chromatic, texture or matter iterations through several sketches. Awareness of emotions in relation to the symbolic matter, history and visual semiotics of the analysed Neolithic culture and writing them on pieces of paper allowed each student to enrich her personal artistic reference. Thus, it was observed in the iteration sketches that one student was more impressed by the structure of Neolithic matter, another by its semiotics and the third one by the texture of matter created by Vadastra Neolithic objects, in relation to the texture of archaic Romanati costumes and tapestry (image 3).

3.	Discussions regarding the direction and meaning given by the impression sketches and the elaboration of the idea sketches for outlining a decorative artistic object. Awareness of personal emotional impression allowed textile arts students to understand that these feelings are the basic elements of their own creativity. Having written them down on paper, they could more easily elaborate idea sketches for different types of decorative art objects. Thus, some thought of monumental tapestries, others of textile prints for fashion or of tapestries for the home ambient (image 4).

4.	Conceptualization of the object, through successive stages of practice of "design thinking", making, remaking, redesigning the concepts until a sustainable sketch is finalized. The sketches conceptualizing the modern textile art objects, within this project, were not of the same quality level as the initial sketches because the students were working on three projects at the same time: the semester one, the bachelor's degree and the research one. As it was not possible to highlight a continuity of the conceptualization process, the team decided to select an existing sketch for layout, from the initial stage (image 5).

5.	Selecting the sketch that can be transposed into a textile art object. Between the expression of the symbol, the clay, the signs and the texture, the artistic

[8] Romania interbelica, 2000 http://romaniainterbelica.memoria.ro/judete/romanati/, http://www.muzeultaranuluiroman.ro/acasa/costumul-traditional-in-romania-ro.html.

practice research team chose a sketch that reproduces the Neolithic texture because it was shown that it has the emotional charge necessary to understand that historical period. The student whose sketch is transposed in the subject wanted to reproduce the Neolithic archaism of the texture and its emotion in relation to this 5000-year-old world. Emotion transgresses the image (image 6).

6. Layout in the handloom after material tests, weaving tests and choosing the optimal relational variant: material/creative weaving technique/archaic expression, Neolithic/biomaterial/modernity (image 7). The layout had two directions: the direction of physical simulation of the Neolithic weaving with the preservation of the ancestral elements, and the direction of rendering the same sketch by correct haut-lisse weaving. The layout in archaic textile material, undyed and dyed hemp, used both for warp and weft, by reconstructing the technique of vertical weaving, also received an archaeological value from the student, namely, of time wear. The result is a model of a modern tapestry that can be exhibited, although it is destructured, but which conveys the idea of an archaic texture. The second model is of a modern tapestry, with an abstract composition, which transmits only the appropriateness of a correctly executed textile work (image 8).

7. The virtual layout of three variants of the basic sketch, generating the Neolithic texture, led to the proposal of some tapestries for interior design: ethnic, classic, elitist. The layout through VR programs has shown that all three upholstery proposals for the public space, a Romanian-style boutique hotel (image 9), an institutional space (image 10) and a modern home (image 11) are proposals that can be put into practice because they have the necessary aesthetic and commercial values of a trade with decorative, modern and civilized art objects.

4   Conclusions

The research project in the practice of textile art, "VADASTRA—Self-referentiality for the re-crafting design" demonstrated that the artist's self-referentiality in relation to the cultural and physical matter of the research topic can lead to a multitude of culturally sustainable solutions if the material and procedural matrix are sustainable.

The project verified that the "design thinking" method understood in its details can be a useful method in the sustainable development of decorative art objects.

Through the mix of information, specific to a complex documentary research, as well as to the imaginary and emotional explorations of the textile matter, the artistic self-referentiality is stimulated, so that the meaning and direction of the research will not be lost, and the artistic expression will be enriched by knowledge.

The exploratory and experimental approach of the archaic textile culture will determine the fixation of the theoretical knowledge but also the sensitization of the artistic self and of the cultural memory which will allow the creation of sustainable identity products in any field of re-crafting design.

Image 2 Romanati tapestries and ethnographic costume, nineteenth century, collections of the Museum of the Romanian Peasant, Bucharest, Romania http://www.muzeultaranuluiroman.ro/acasa/costumul-traditional-in-romania-ro.html

Image 3 Sketches of empathy and iteration based on emotional immersion and students' study in material and historical culture Vadastra and Romanati (The image of the author, with the consent of the students)

Simulating the display of textile art objects in their living environment, through Virtual Reality applications, is a form of educating the aesthetic taste of both designers and future clients, by posting the projects on social media platforms of specialized trade.

Image 4 Part of the research team, in project discussions: Laura Eftimie and Claudia Davidescu—third year students, bachelor, Textile Arts Department, UNA Bucharest, Romania (The image of the author, with the consent of the students)

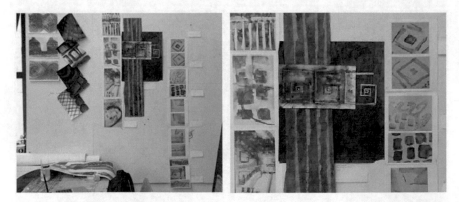

Image 5 Conceptualization of the object, through successive stages of practice of "design thinking", making, remaking, redesigning the concepts until a sustainable sketch is finalized. (The image of the author, with the consent of the students)

The method of cultural self-referentiality in the textile arts, corroborated with the application of the method of design-thinking in the decorative arts, demonstrated in this experiment that there is a dialogical compatibility that can be developed in artistic language. This fact will lead in the future to the dialogical deepening of these methods, the development of design thinking techniques specific to decorative arts through other exploratory projects in art practice research.

Research in art practice has shown that, the cultural sustainability of a textile decorative art object is the synergistic effect of the relation between the cultural identity heritage value of the artistic or design theme approached by the artist, cultural values of the author, brand, ethnic or nation and the use of the aesthetic and ecological material, supports necessary to express it in a new product concept, innovative and modern.

Image 6 Selecting the sketch that can be transposed into a textile art object. Research team chose a sketch that reproduces the Neolithic texture because it was shown that it has the emotional charge necessary to understand that historical period. (The image of the author, with the consent of the students)

Image 7 Tests of materials and weaving to make models of textile art objects (The image of the author, with the consent of the students)

Image 8 Layout of the sketch of ideas chosen by the team, in two variants: the empathic reproduction variant of the Neolithic and the traditional tissue variant, both models were made on the vertical frame, with hemp threads. (The image of the author, with the consent of the students)

Image 9 VR simulation, to suggest the integration of the tapestry within a Romanian ethnic style, in a boutique-hotel in Romanian style (The image of the author)

Image 10 VR simulation, to suggest the integration of the tapestry within a public space (The image of the author)

Image 11   VR simulation to suggest the integration of tapestry (The image of the author)

References

1. Anderson J, Shattuck L (2012) Design-based research: a decade of progress in education research? Educ Res nr 41(2012):16–25
2. Gheorghiu D (2001) Le projet Vădastra, Prehistorie Europeenne. Liege, pp 16–17
3. Gheorghiu D (2008) Cultural landscapes in the Lower Danube area. Experimenting tell settlements, Documenta Praehistorica XXXV. Liege, pp. 7–13
4. Gheorghiu D, Rusu A (2008) Cultural landscapes in the Lower Danube area. Experimenting tell settlements, Documenta Praehistorica XXXV. Liege, pp 7–13
5. Kimbell L (2011) Rethinking design thinking. Part I, Design Cult nr 3(2011):285–330
6. Jin Y, Chusilp P (2006) Study of mental iteration in different design situations. Des Stud 27(1):25–55. https://doi.org/10.1016/j.destud.2005.06.003
7. Lawson H (1985) Reflexivity: the post-modern predicament. Hutchinson, London
8. Negru M, Bădescu Al, Avram R (2006) Pottery of terra sigillata type discovered In Roman forts of Dacia (Inferior) Malvensis. Published by MUZEUL NAȚIONAL DE ISTORIE A ROMÂNIEI in the Cercetări arheologice, no XIII, pp 231–238
9. Lawson, Noth (2001) Rethinking design thinking. Part I, Design and culture, vol 3, pp 285-230
10. Lotman, YM (1990) Universe of the mind: a semiotic theory of culture, Ann Shukman (trans.). I. B. Tauris, London
11. Marcus S (1997) Media and self-reference: the forgotten initial state. In: Nöth W (ed) Semiotics of the media: state of the art, projects, and perspectives. Mouton de Gruyter, Berlin, pp 15–45
12. Noth W (2006) Representations of imaginary, nonexistent, or nonfigurative objects. Cognitio 7(2). 247–259. Sao Paulo
13. Nöth W (2001) Autorreferencialidad en la crisis de la modernidad. Cuadernos 17:365–369
14. Nöth W, Ljungberg C (2003) The crisis of representation: semiotic foundations and manifestations in culture and the media. Spec Issue, Semiotica 143(1/4)

15. Moreno DP, Hernandez AA, Yang MC, Otto KN, Linsey JS, Wood KL, Linden A (2014) Fundamental studies in design-by-analogy: a focus on domain-knowledge experts and applications to transactional design problems. Design Stud 35(3):232–272. Elsevier
16. Hekkert P (2006) Design aesthetics: principles of pleasure in design. Psychol Sci 48(2): 157–172
17. O'Sullivan T et al (1994) Key concepts in communication and cultural studies. Routledge, London
18. Pop M, Iuhas F, Borangic C (2016) Repertoriul imagistic Roman, Ed incdtp-icpi Bucuresti, ISBN 978-973-0-23249
19. Pop M, Toma S, Frigy N (2018) Cultural work instruments in fashion technology - practical method of generating emotional design, J Textile Sci Fashion Tech 1(1):2018. JTSFT.MS.ID. 000505, San Francisco, USA
20. Weber C (2005) International relations theory, a critical introduction, simultaneously published in the USA and Canada by Routledge 270 Madison Ave, New York, NY 10016 Routledge is an imprint of the Taylor & Francis Group, ISBN 0-203-48146-1
21. https://www.collinsdictionary.com
22. https://www.mnir.ro/
23. http://www.muzeultaranuluiroman.ro/acasa/costumul-traditional-in-romania-ro.html, http://romaniainterbelica.memoria.ro/judete/romanati/
24. https://www.youtube.com/watch?v=J8MKr_NfGHA&t=662s   https://www.youtube.com/watch?v=HAldsF6SCyE&t=742s

Conscious, Collaborative Clothing: A Case Study on Regenerating Relationships Within the Khadi Value Chain

Ashna Patel

Abstract Globalisation and industrialisation has led to the social relations involved in cultivating our everyday products, such as clothing and textiles, to become invisible, resulting in inequity and vulnerability within value chains. These complex, globalised value chains have led to a lack of connection between our actions and the corresponding impacts on both human and non-human ecosystems. This case study addresses and challenges this disconnection by setting out to regenerate relationships within the clothing and textile value chain. The study draws on existing research in the field of design for sustainability and ethnographic research conducted in England, in collaboration with activists and suppliers of khadi, a natural fabric originating from India, traditionally handwoven with handspun yarn on a handloom. The chapter discusses the birth of the khadi movement, its socio-economic and environmental impacts and its development in the khadi sector. A collaboration is also developed between the designer and members of a local community in Kolding, Denmark, with the aim of gaining a deeper insight into relationships with clothing and factors that influence fulfilment and longevity. The workshops simultaneously acted as a bridge to share knowledge about the significance of natural fibres further down the value chain and led to co-designing a limited series of artefacts. The goal of the case study is to investigate if and how building upon collaborative relationships within the value chain has the capacity to challenge the current speed and scale of design, production and consumer culture, and to pose an alternative practice that aligns with planetary boundaries and human well-being. Thus, opening up a path for designers to consider honing in on developing hand-crafted luxury garments that align with principles of environmental and social sustainability in the field of clothing and textiles.

Keywords Khadi · Handloom · Sustainable design · Slow fashion · Natural fibres · Collaboration · Value chain

A. Patel (✉)
Design School Kolding, Kolding, Denmark
e-mail: ashna_cpatel@hotmail.com

© The Author(s), under exclusive license to Springer Nature Singapore Pte Ltd. 2021   135
M. Á Gardetti and S. S. Muthu (eds.), *Handloom Sustainability and Culture*,
Sustainable Textiles: Production, Processing, Manufacturing & Chemistry,
https://doi.org/10.1007/978-981-16-5967-6_7

1 Introduction

The era of industrialisation gave rise to the globalisation and complexity of value chains. Prior to this, products were designed and made by local craftspeople—often utilising local resources—as a response to the needs within a community, and the relationships between products, human labour and nature were visible. This, however, is no longer the case. Instead, industrialisation has transformed clothing and textile production from a decentralised and diverse system into a centralised system [1] where we are no longer exposed to or connected with these relationships. As a result, this has led to extensive exploitation, inequity and extraction of our human and non-human ecosystems in favour of a system of 'conspicuous over-production known as 'fast fashion' in which low prices feed and enable over-consumption' [8: 17]. Our separation from the human and planetary resources that are responsible for the everyday products we use, such as clothing and textiles, has resulted in apathy towards the source, and the illusion of the 'other' [36: 114–115]. This reality in which we centre and prioritise ourselves is reflected in our current design, production and consumption culture.

However, there is hope for the slow and steady, but impactful movement of more responsible design practice to proliferate; one which lends the opportunity to challenge the extractive mainstream business model. Design for sustainability practices hold the potential to be agents for positive change; to include communities, exchange knowledge, preserve and innovate tradition, create fair livelihoods and allow space for sharing and storytelling what we learn along the way. Fletcher and Tham [8: 40] explain, 'this work is about very real world problems, real world solutions, to be enacted by real world people.' Their action plan addresses replacing Growth Logic (priority: money) with Earth Logic (priority: Mother Earth and its inhabitants). Instead of simply denouncing current paradigms, they offer alternative strategies of fashion being inclusive and experiential, losing our usual 'me, myself and I' mentality to instead engage in symbiotic relationships in which we consider and acknowledge the 'other' [36: 114–115]. It touches upon examples of how designers can reimagine their practices beyond the walls of their studios, through, for example, deeper industry and interdisciplinary collaboration, reconnection with nature and inclusivity of our surrounding communities.

The case study takes inspiration from these emerging approaches. It is important to note that the authors of *Earth Logic Fashion Action Research Plan* reassure us that feelings of uncertainty and discomfort while exploring these possibilities may surface. They describe systemic change as being 'non-innocent', and reference Donna Haraway's guidance to 'get closer—and stay with – 'the trouble" and to commit to 'the difficult, uncompromising task of trying to live better together on a damaged planet' [8: 7]. This advice speaks to the many trials and tribulations faced along the way. The study is heavily influenced by imagining new ways that clothing and textiles can exist in a world that prioritises connection to Earth, each other and ourselves—a regenerative economy. In a seminar hosted by Fibershed, enti-tled *Regenerating Our Textile Systems: Defining Regeneration for Fiber and Textile Systems* [3], Dayaneni defines a regenerative economy aptly in this way:

> The economy we need, the one we have to remember our way back into, is one in which the purpose is social and ecological well-being. In a regenerative economy the purpose is not to maximise profit but actually to meet people's textile needs in such a way that is sustained over time and resilient. And to do that you have to have a different relationship with the living world and human labour—we think of that as corporation and regeneration [15].

The meta context of the study lies within the principles of a regenerative economy and proposes an alternative methodology of designing and producing garments *with* regenerative resources and *through* creative democracy. The case study aims to culti-vate visibility of the social relations within clothing value chains and investigates the impacts that building upon these social relations can have on clothing design practice. It explores how forming meaningful stakeholder relationships may initiate a more self-reflexive practice which questions, for example, the textiles that we incorporate into our design practices and the ways in which those textiles are sourced. Many modern textiles have often been manipulated to prioritise convenience and mass production at the expense of the health and well-being of our planetary and human resources. Due to production-to-consumption systems being spread across nations and geographies, knowledge on and reliability of such systems often lacks trans-parency and results in the opportunities of design intervention becoming challenging [28: 44]. Thus, gaining a deeper understanding through craft-design collaboration may overcome barriers to a holistically sustainable design practice [28: 44].

After a brief discussion on the methodology in the next section, the third section introduces the collaborative relationship with a UK-based social enterprise whose primary goal is to promote and supply khadi, a handspun, handwoven natural fabric originating in India, while helping to sustain the livelihoods of rural farmers and artisans, and preserve the tradition of khadi. The collaboration took the form of exchanging knowledge and resources on the birth of the khadi movement, its socio-economic impacts, and development in the khadi sector which are outlined in the fourth section of the chapter. In the following section, the focus shifts to how this knowledge may translate into, for example, a more meaningful design practice, and a contribution to enhancing the potential of khadi, thus supporting rural livelihoods. In section six, the chapter goes on to dissect an investigation on how designers might step into the role of facilitators to work with wearers through workshops in order to challenge and propose alternatives to unsustainable clothing behaviours. Section seven centres around the impacts and potential that building bridges between stakeholders may have in aligning with holistically sustainable design practice, and the tangible outcome of khadi artefacts are presented. Finally, the conclusions of this case study are drawn.

2   Methodology

The case study employed a range of design and science research approaches in order to obtain an overview of the contribution of collaborative design practice to holistic sustainability within the design. The study was grounded in Kolding, Denmark, where collaboration with wearers of clothing took place, and it also branched out to collaborate with a natural-fibre-focussed social enterprise based in London, England, which have strong links to khadi initiatives in rural India. These branches opened up nuanced perspectives that were both local and global. Due to the diverse nature of the roles these collaboration partners have in the value chain, different methods of design and science were employed when engaging with each group and combined theory, practice, experience and processes of articulation [8: 23]. The science methodologies included case study, desk-based and field study, and the design research methodologies combined ethnography, practice-based and workshop development and facilitation. Action research, detailed in Earth Logic as involving synergistic research and change-making which researches with rather than on people, was a source of inspiration throughout the investigation [8: 23]. The inclusion of stakeholders into a typically solitary design process required cycles of action and reflection to process the variety and quantity of information gathered into actionable insights [8: 15]. The case study resulted in a limited series of six co-designed artefacts that narrate an iterative process and insights collected along the way. The case study aimed to bridge the gap between an ethical khadi supplier, designer, and wearers, and in doing so, asked the following questions:

- How might designers work alongside textile initiatives that are rooted in natural fibres, tradition and the well-being of their value chain?
- How might designer-wearer relationships develop to cultivate garments that meet the fundamental needs of wearers?

3   In Collaboration with a Social Enterprise: Khadi London

Khadi London is a Community-Interest-Company based in England, U.K. They source and distribute material that is responsibly produced by co-operative and community-based farmers, graziers and artisan clusters in India. They exist to promote the use of khadi in the fashion and textile industry and to 'support the work of designers who value craft and provenance while challenging the unsustainable practices of fast fashion and the throwaway culture' [13].

To learn more about Khadi London and khadi itself, a visit was organised to carry out ethnographic design research for the period of one week. Shadowing of the founder of Khadi London, Kishore Shah, was employed as the primary methodology alongside semi-structured interviews and conversations that were loose and intuitive in nature, to allow for both stabilisation and drift [27]. As Reubens highlights, 'if the business does not incorporate sustainability at a strategic level, sustainability concerns do not generally trickle down to its key business systems–including design' [28: 44]. The visit was vital in order to determine whether the inner-workings, ethics and goals of Khadi London, showed potential to contribute to a holistically sustainable design practice. The field study strengthened Khadi London's role as a natural fibre activist part of a larger movement towards positive global change within the fashion and textile industry. Since it was founded in 2014, the company has branched out to form non-transactional relationships with producers and customers to better understand how they can be an ally for social and environmental justice. They have an innate understanding that in order to thrive, their value chain needs to function in symbiosis with other members. They support this by often shifting into the role of facilitators bridging the gap between khadi producers and customers to open up the potential of self-initiated collaborations, standing by to support and ensure that equitable relationships are developed. Furthermore, in collaboration with a growing network of ethical and eco-conscious stakeholders, Khadi London has co-hosted a yearly event, 'The Festival of Natural Fibres'. The festival invites all stakeholders to participate in workshops, presentations, panel discussions and open dialogues regarding, for example, their contributions and practices, how they navigate and resolve the challenges they face, and their experiences of working with members across their value chains whether on a local or global scale. Twigger Holroyd discusses that there are many initiatives seeking to revitalise culturally significant designs, products or practices–often in collaboration with designers [12: 25]–and goes on to define this role as 'any initiative that brings new life to a culturally significant design, product or practice, while aiming to retain (or even enhance) the values associated with it' [12: 30]. The Festival of Natural Fibres is just one example of how Khadi London aims to fulfil this mission by engaging with an array of stakeholders to share and promote the movement, message and values of khadi amongst eco-conscious design and craft communities. In doing so, stakeholders are offered the opportunity to employ khadi in their practices to creatively interact within the space between tradition and innovation. This case study is an example of this opportunity in practice.

Taking inspiration from Menon and Uzramma's investigation of the journey of cotton in India [24], Khadi London are led by a 3D principle–diverse, democratic, decentralised–ensuring that the fabrics they offer are cultivated in symbiosis with their ecosystem, alongside both human and non-human actors: 'The idea of a khadi economy is an economy guided by the concept of diverse, democratic and decentralised production systems. Systems which respect the environment and embrace principles of fairness and social justice' [13].

Diverse: Promoting diversity in seeds, breeds, other raw materials and production methods, bringing together the best in local traditions and modern technology.

Democratic: Aiming to give maximum control over ownership to farmers, graziers, artisans, communities and cooperatives.

Decentralised: Production through a network of decentralised units within a cluster/region with the maximum value addition at the source.

To Khadi London and initiatives alike, khadi is not just a material; it is a philosophy, an ideology and a movement.

4   An Introduction to Khadi

4.1   Cotton to Cloth

Khadi is a handspun, handwoven nature fibre originating from India. Traditionally, khadi is made from cotton, but there are also varieties of khadi wool and khadi silk depending on the region where it is spun. Khadi distinguishes itself from being categorised as simply cotton, silk or wool through the processes of hand spinning the yarn on a charkha (spinning wheel) and then weaving the handspun yarn on a handloom.

The khadi value chain consists of five critical stages: cultivation of cotton, ginning, carding, spinning and the production of fabrics/textiles [18].

Cultivation of cotton: Cotton seeds are planted and then harvested over several months. The varieties of the cotton seeds depend on each region's ecology (Fig. 1).

Ginning: A process which separates fibres from seeds. Traditionally, ginning is done by hand, but many khadi producers and initiatives now utilise mechanised ginning machines.

Carding: A process which cleans and detangles the fibres to produce continuous slivers which are then ready to be spun in the next stage. Again, traditionally carding was carried out through a hand-powered machine, but developments have led to mechanised carding machines.

Fig. 1 Cotton seeds, cotton bolls, and husks at khadi unit, Gram Seva Mandal, founded by Vinoba Bhave in 1934 in Wardha, Maharashtra, India. *Source* Project PICO—Photograph by Project PICO. Published with the authorisation of Project PICO

Fig. 2 A skilled spinner hand spinning on a traditional charkha (spinning wheel). *Source* Khadi London—Photograph by Aaron Sinift. Published with the authorisation of Aaron Sinift [33]

Spinning: The hand spinning of the yarn is what distinguishes khadi from cotton. Traditionally, the spinning of the cotton is done on a charkha, allowing for only one sliver of yarn to be spun at a time. Khadi units have since developed ambar charkhas, which allow for multiple slivers of cotton yarn to be spun at a time. Evenly spun and twisted yarn produces a softer, more durable khadi cloth (Fig. 2).

Production of fabrics/textiles by craftspeople, including yarn procurement, yarn dyeing and weaving: Once the cotton fibres have been spun to produce a yarn, they go through a warping process before being ready to be woven on a handloom, or power-loom, which emerged in the 1920s 'as an intermediary between the pre-existing hand-loom and mill segments of the textile industry' [25: 6]. The final cloth is then ready to have textile techniques applied to its surface, including dyeing, printing, embroidering, etc. The textile techniques applied throughout khadi production depend on the region's know-how, with each region in India, specialising in their own textile crafts that have often travelled through generations (Figs. 3, 4 and 5).

Fig. 3 Naturally dyed yarns hanging to dry. *Source* Henri London—Photograph by Henri London. Published with the authorisation of Henri London

Fig. 4 Creating the warp to be woven on a handloom. *Source* Khadi London—Photograph by Susanta Biswas. Published with the authorisation of Susanta Biswas/Khadi London

Fig. 5 A weaver preparing the warp on a handloom. *Source* Henri London—Photograph by Henri London. Published with the authorisation of Henri London

4.2   A Brief History

Khadi remains a culturally significant material imbued with social, historical, aesthetic [12: 27], and spiritual values, and a symbol of India's struggle for independence. Khadi was built upon the Swadeshi movement between 1906 and 1911, 'which sought to establish India's economic autonomy from Britain as a basis of self-government' [38: 17]. The khadi movement was born from Mahatma Gandhi's nonviolent protest to boycott the import and cultivation of factory-made textiles and clothing that were manufactured by British companies. As Reubens quotes Khan, 'thriving craft-based export industries—such as India's handloom sector— were systematically sabotaged by colonial policies designed to reduce exports, while simultaneously leveraging the colonies as lucrative markets, which could absorb industrialised imports' [28: 50]. The Swadeshi movement supported handloom weavers who were struggling to survive due to the competition from industrial textile mills. Khadi at this point had more or less become extinct. In Gandhi's autobiography he proclaims:

> The object that we set before ourselves was to be able to clothe ourselves entirely in cloth manufactured by our own hands. We therefore forthwith discarded the use of mill-woven cloth, and all the members of the Ashram resolved to wear hand-woven cloth made from Indian yarn only. The adoption of this practice brought us a world of experience. It enabled us to know, from direct contact, the conditions of life among the weavers, the extent of their

production, the handicaps in the way of their obtaining their yarn supply, the way in which they were being made victims of fraud, and, lastly, their ever growing indebtedness. We were not in a position immediately to manufacture all the cloth for our needs [10: 539].

This realisation is when the importance of yarn cultivation became evident, and thus led to a sense of urgency for the revival of hand spinning. The search for charkhas and artisans who could impart the skill of spinning began and was achieved within a few years, with viable spinning units set up in a village in Gujarat (Vijapur) and at Gandhi's ashram. It was proved that cloth production could once again become independent of textile mills, and it was now only a matter of scaling. The khadi movement was born, to eventually become a major driver for India's freedom struggle [31].

Vinoba Bhave, a close ally of Gandhi for about thirty years, was offered a hundred acres of land for landless labourers. It was a spontaneous offer aimed at bringing an end to a violent conflict over land in the village. It was followed up by similar offers in neighbouring areas—a small act of land gift soon became a movement. Gandhi's principles of nonviolence proved to also work for resolving internal disputes, and a spirit of reconciliation showed promise in eventually replacing the need for a state dependent on the rule of law. Gandhi's loyalists found a new energy as their message resonated with the rural masses [31].

Bhoodan, as the land gift movement was called, had wider impacts. Cottage industries were added as a niche area for planning which in turn enabled the craft of khadi to survive. In 1957, a law was enacted to establish a statutory body, Khadi and Village Industries Commission (KVIC), to support khadi and other rural industries. The movement continued to have a significant political influence for about twenty-five years during which rural industries have thrived [31]. According to a report by Khadi and Village Industries Commission, khadi alone gives employment to half a million, about 80% of whom are women [14]—a large number in remote rural areas. New forces, including the rise and awareness of sustainable design, continue to drive khadi forward.

4.3 Handloom and Khadi

Both khadi and handloom fabrics were boosted with the revival of khadi in the 1920s due to the important role khadi played as part of India's freedom struggle, especially in rural India where it had the added advantage of becoming a source of livelihood. Khadi and handloom, however, had very different trajectories for growth. Khadi in many ways pioneered the spirit of social enterprise, whereas handloom was driven more by tradition and a capitalist ethos. Handloom, when viewed from a purely market perspective, has a single focus on textiles. For example, handloom production is supported by the Textile Ministry, while khadi in contrast, by the Ministry of Small and Medium Enterprises. Handloom is viewed from a craft perspective and promoted as such, and khadi as a rural enterprise, providing a model for resilient economies and a pivot for non-farm activities. Khadi institutions also tend to be rooted in the Gandhian tradition where Gram Swaraj (village autonomy) forms the basis for a strong economy and polity at the national level. For idealists who believe in Gram Swaraj, khadi is part of a larger whole, including other rural crafts and industries, sustainable agricultural practices, and perhaps more importantly, the ideas of participatory democracy at the village level, of social justice and fairness, and of equity, including a strong focus on gender equity [31].

4.4 Social, Environmental and Economic Impacts

The aim of khadi production was not to create a profit-making business, rather cultivate a socio-economic activity to render self-sufficiency, earning opportunities and welfare of rural artisans [34], at little to no environmental cost. As khadi production works within localised production systems, farmers and artisans maintain a strong sense of ecological stewardship towards their respective bioregions [28: 52].

Post-independence, India saw a transition from the decentralisation and diversity of yarn and textile production to centralised large-scale industrialisation which led to mill-made yarn and textiles that provided uniformity of raw material [1], something that the hand-processes of khadi production does not achieve. Relationships in the value chain between, for example, farmers, spinners and weavers, that once were equal, shifted in favour of the mechanised spinning mills [24]. Not only were they able to dictate the financial terms of their exchange, but also the quality of the yarns. The machinery in the mills preferred long-staple fibres that mill-spun yarn was able to offer, whereas handspun yarn tended to be short-staple. In order to sustain their livelihoods, farmers had to resort to using foreign seeds—as opposed to indigenous seed varieties which they were familiar with—which offered longer-staple fibres. However, the foreign varieties had a low resistance towards tropical climate borne diseases and pest attacks as they were not native to the Indian subcontinent [9]. Farmers had no choice but to shift from indigenous farming methods, which were inherently organic, and invest in pesticides to ensure their cotton crops would survive. However, the investments into foreign seed varieties and pesticides proved to work against the farmers' financial stability, instead facing them with debts. The 1990s saw a surge in farmer suicides.

Currently, several production groups, initiatives and organisations in India are looking more closely at the value chain to support khadi producers including farmers, spinners and weavers in reclaiming stable livelihoods by, for example, shifting back to indigenous farming practices including reintroducing indigenous and local seed varieties (as opposed to hybrid and/or genetically modified seeds) and abandoning the use of pesticides [24].

> Khadi has a wider vision. The soul of this unique system is that it is based on very humane and inclusive ethics and morality. Clothing is indeed one of our primary requirements. Khadi brings the ethics back to clothing [32].

4.5 Developments in the Khadi Sector

Spindle to Solar

In many cases, spinners are paid according to the amount of yarn they are able to produce [31]. The hand-powered nature of the spinning process means that there is pressure on the spinners to work at a fast pace if they are to make a decent wage. The development of the ambar chakra, which ranges from 2 to 12 spindles, as opposed to just one which a traditional charkha offers, enables a more economically viable solution. Recent developments have seen a move towards solar-powered ambar charkhas which would increase productivity and thus the khadi economy. India's clear and sunny weather conditions have the ability to produce the solar energy of '5000 trillion kilowatt-hours (kWh) per year (or 5 EWh/yr) [exceeding] the possible energy output of all fossil fuel energy reserves in India' [32]. Thus, khadi production has the potential to become a minimum or zero-carbon emission process.

Khadi clusters

The idea of a khadi economy presents an opportunity for cluster-based initiatives that are able to recreate the farm-to-fabric production system within their own regions. There are many interchangeable factors depending on the region's human and environmental resources, such as the natural fibre that is suited to the region's climate and ecosystem. However, what is retained is the concept of a community-cultivated cloth. This concept is explored by Fibershed, an organisation that investigates the decentralisation of textile processes led by regional textile communities in different locations around the globe [2]. For both khadi initiatives and Fibershed, the goal is to move towards self-sufficient local economies whereby fibres are grown, harvested, processed, woven and cultivated into clothing and textiles that fulfil fundamental human needs, and at the end of their life, can safely return back to the soil. Kishore Shah, the founder of Khadi London, writes about an event that took place as an extension of London Fashion Week in 2017 involving a multitude of natural fibre stakeholders: 'The conference saw a shift in our collective thinking, from khadi being a heritage fabric made in India to cluster-based production. The idea of khadi had a role to play in the UK using UK-based natural fibres such as wool, hemp, flax and nettle. The same model can be transferred anywhere in the globe, using local natural fibres, and the region's own socio-cultural skill sets and resources' [31].

Preservation of craft

The readily available, abundant and cheaper alternatives to handmade products pose significant challenges to the survival of traditional forms of craftsmanship [37]. Our connection to craft and the human labour involved has slowly disappeared as products of globalisation and industrialisation have infiltrated mainstream markets globally. Reubens addresses a post-2015 UN development agenda which call for sustainable development to be inclusive of economic growth, decent employment, social justice and protection, and environmental stewardship and goes on to say that craft, such as khadi production, holds the ability to 'achieve all of these holistically, and thus impact all four tenets of sustainability including ecological sustainability, social sustainability, economic sustainability and cultural sustainability' [28: 52]. Khadi production also contributes greatly to the transmission of generational knowledge and intangible cultural heritage such as hand spinning and weaving that would otherwise become lost crafts. As a shift in awareness amongst conscious creators and consumers grows, as does crafts' opportunity to be a vehicle of holistic sustainability.

Aesthetically, craft-based textiles offer a softer alternative to perfectly uniform textile surfaces created by machines, and are instead infused with the spirited qualities of handmade textiles. The handmade process of khadi often results in inconsistencies and flaws in the yarn and final cloth, such as slub yarn (a yarn with thick and thin sections alternating regularly or irregularly), mottling and/or cotton seeds visible on the fabric's surface: 'such artefacts often touch people more deeply than machine made goods because the mark of the human hand can create a connection through form. We understand it, relate to it and are inclined to value it. Quite simply, it matters to us' [42].

5   Opportunities of Collaboration with Khadi London

Through a combination of ethnographic and desk-based research, and semi-structured interviews, knowledge developed on khadi as a natural fibre, and the foundation, values and goals of Khadi London. The following interconnected opportunities presented themselves as focus areas of the collaborative case study.

- Evolution of khadi's aesthetic language

Khadi London continually works towards promoting the khadi movement to larger networks to stabilise and sustain equitable livelihoods within their value chain. Amongst craft communities, especially those in developing countries, though there is an abundance of know-how, there is often a lack of know-what in terms of applying their knowledge and skills to design within the 'complex economic and social matrix in which they exist' [26: 14]. Utilising design skills and competences to evolve the visual language of khadi presented itself as an opportunity in joining Khadi London on their journey to introduce and/or open up the potential of khadi to a wider network of stakeholders.

- Living into a meaningful relationship with an integral member of the value chain

In order to move towards the role of a decentralised designer to see what opportunities and contributions this could bring to sustainable design practice, a meaningful collaboration was developed through the sharing of knowledge, networks, competences and experiences. The aim of the collaboration was to reach a place of shared potential and opportunities.

- Offering a new reality of the design and production of clothing

Khadi London shared their interest in establishing a business model that is more meaningful than current industry norms, prioritising the creation of sustainable livelihoods that support the preservation of tradition, culture and craft, and abandoning the goal of economic growth. This case study offered a way in which Khadi London might pair their philosophy and ethically cultivated fabrics with a conscious design and production process.

- Extending material knowledge further on in the value chain to instigate behavioural change

As the collaboration developed, knowledge about khadi's values, cultivation, impacts on human and non-human ecosystems deepened and presented khadi as a more responsible fabric choice for clothing design practice. This new-found material knowledge could then be passed down further on in the value chain to wearers through workshops, in order to instigate reflection and reassessment of mainstream consumer behaviour.

6 Development of Khadi Artefacts

The case study takes the form of researching on and through material objects and things. Woodward distinguishes these terms as objects being entities in themselves and things existing in context, as relations [43]. Artefacts in the form of garments were developed and utilised as a basis for collaborative investigation. The cultivation of artefacts enabled khadi to be translated from a cloth to an object with shape and form that increases in the way people relate to it.

> Clothing is so much more than just "fashion"–it sends important signals about our place in society, both actual and perceived, and is a critical element of our personal narrative—the daily, even hourly, choices we make to communicate our histories, desires, affiliations, and self-image... [2: 17] The idea of being warm when you need to be warm, cool when you need to be cool, protected from sun, and having clothing that moves well with your movements, clothing that honours a commemorative event—these are the fundamental reasons we have clothes. [2: 58].

Great importance was placed on exploring the intersection and entanglement of clothing and lifestyles. In this regard, the development of shapes and forms are grounded in the vision of 'everyday' garments that balance comfort and purpose with beauty and simplicity, and that hang beautifully on the body. The shapes were simplified to leave space for wearer engagement in the following phases and to allow room for khadi's own visual language to take precedence. A process of designing, developing, co-designing with wearers and prototyping resulted in six final artefacts cultivated from khadi cotton supplied by Khadi London and their khadi producers in rural India.

From a wider perspective, the development of artefacts stands to exemplify ideas and visions for alternative methodologies of practice: 'Although designers produce design objects, these can also be treated as experimental artefacts that produce knowledge' [19: 50]. This study proposes that garments can be cultivated as 'cultural probes' [11] with multi-faceted research possibilities including but not limited to: aesthetics of clothing, wearer experiences, opportunities for sustainable development and new realities and economies of cultivating clothing.

6.1　Exploring a Diverse, Democratic and Decentralised Design Practice

Tham and Jones define the concept of 'metadesign' as an idea that 'advocates design that operates at systemic levels, that invites interdisciplinary collaborations and that seeds or sets up the conditions for emergent processes of change' [35]. The case study took inspiration from this idea and in doing so, invited community members from Kolding, Denmark into a design collaboration. The idea of designer-wearer collaboration was strengthened by Khadi London's 3D (diverse, democratic, decentralised) vision. For them, this framework is used to guide the organisation of material processing and distribution, however, the aim, in this case, was to transfer the framework to garment design and production by exploring a method of practice that is inclusive and non-hierarchical [35].

The primary goals of forming relationships with wearers of clothing were to open up dialogues around clothing and our relationships to clothing that are often very private, to understand and discover nuances around wearers' needs in order to profoundly understand clothing relationships and how to offer emotionally, technically and aesthetically durable designs [4]. This may lead to opportunities for design intervention for, for example, the fibres that make up a garment and speculating new and sustainable realities of clothing experiences.

In order to test how elements of a more collaborative practice would work, look and feel, design tools and workshops were designed to explore the idea of living into 'nurturing, fostering, and harnessing' [35] designer-wearer synergy in the value chain in order to discover opportunities of shifting our current design and consumer culture. The collaboration with wearers began with digital and physical wardrobe studies conducted both individually by wearers' and with the facilitation of the designer [16]. Iterative reflections and insights guided the case study towards a community workshop where wearers were presented with six simple garment prototypes and an archive of prototyped design features including a variety of collars, waistbands, fastenings, methods of adjustability, etc. The aim was to receive engagement and feedback on the garments and design features while encouraging a wider dialogue that centres around their clothing relationships, habits and needs.

The structure of the community workshop was influenced by the four-fold integrative framework developed by Dr. Otto van Nieuwenhuijze, Tham and Jones which aims to 'encourage a more self-reflexive, empathetic, context dependent, inclusive and highly collaborative process than more conventional fragmented and hierarchical design processes' [35]. The structure of the workshop followed the principles laid out in the framework.

Four-fold integrative framework developed by Dr. Otto van Nieuwenhuijze, Tham and Jones [35]	Developed framework for the communnworkshop with wearers
Phase one: Awareness at the level of the self, the designer (My personal involvement, I present myself, what I am bringing to the table)	**Phase one:** Show and tell: the wearers each present 1–2 garments that for them represent the terms: comfort, function, durability and versatility to dissect what these definitions could mean to different wearers. The activity also acts as an icebreaker
Phase two: Awareness at the relational level, 'being with' (Reflection on my personal view, 'sympoiesis' or creating with the other)	**Phase two:** Exploration: wearers are introduced to the prototypes including simple garment shapes and an archive of design features (including collars, waistbands, adjustability features, etc.), and are invited to try them on and explore combinations of garments and design features
Phase three: Awareness at the level of the team (Each individual identifies with the team as a whole, a team 'potential realisation' (Fairclough, 2005) occurs)	**Phase three:** Collective garment reflection: designer and wearers gather to discuss the individual and shared experiences of the garments and to explore a collective conclusion for each garment
Phase four: Awareness at the level of the wider context (The team as a whole identifies with its wider context/s, potential realisation becomes transferable)	**Phase four:** Collective workshop reflection: designer and wearers engage in a reflection session on the experiences of the community workshop, improvements, insights and potentials for a new reality of design and consumer culture

Six members of the community between 20 and 30 years old attended the workshop, some of whom had been engaged in the investigation since the wardrobe study, and others responded to an invitation to the workshop that was posted on a mainstream social media platform. The information gathered during the workshop was rich and diverse, and was a result of 'multiple centres' [8] coming together and sharing space, thoughts and experiences. The community members that attended were from an array of countries (including Canada, Spain, Taiwan, England and Denmark), and each had their individual aesthetic expression, lifestyle, occupation, attitudes, values, and body shape (Fig. 6 and 7).

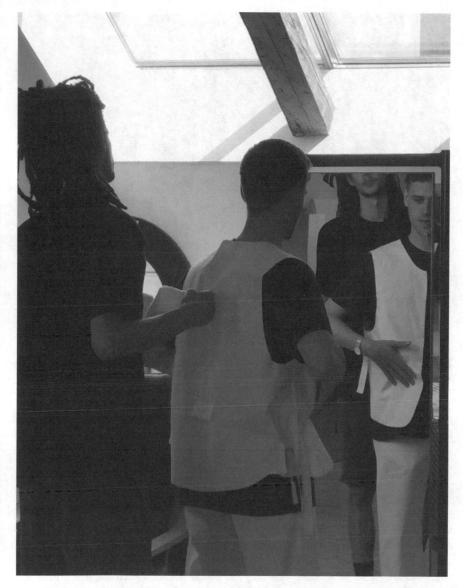

Fig. 6 Two participants discuss the prototypes together during the community workshop. Photograph by Lesley-Ann Donnell

6.2 Response to Khadi

One of the aims of the community workshop was to gain a deeper understanding of wearer relationships with materials, for example, how important the choice of material is in relation to clothing. Another aim was to introduce khadi to the wearers

Fig. 7 The designer and participants discuss the prototypes together during the community workshop. Photograph by Lesley-Ann Donnell

and gauge an initial response. Throughout the workshop, wearers organically referred to materials when trying on the garments:

> I would love to wear this often in both casual places or at home. I guess it depends on the material... If it was a nice, soft linen material, I could be at home with it, go to work with it. I would like to use it daily in multiple situations.

These subtle references brought to light that wearers had an innate awareness of material qualities, both aesthetically and physically. Fabric samples were split into three categories: khadi, medium-heavyweight fabrics including natural canvas, denim and wool, and lighter fabrics such as shirting cotton and linen. Upon interaction, wearers were gravitating towards the khadi fabrics which aligned with many previously referencing the 'breathable' fabrics that they imagine the garments to be made from. Wearers linked the qualities of khadi with garments they could wear every day, underlining that the choice of materials plays a significant role in the purpose of a garment.

6.3 Reflection

Upon reflection, the community workshop revealed that designer-wearer collaboration is an unfamiliar territory that brought excitement and togetherness, as well as moments of confusion and discomfort. Initially, there was some apprehension and uncertainty amongst wearers. However, as they became more comfortable with sharing this usually intimate experience of trying on clothes and verbalising their experiences, the unfamiliarity fuelled excitement and engagement. The wearers were encouraged to explore in a way that felt comfortable to them, whether it be individually or collectively. The reflections from wearers confirmed that (design) researching in this way, can indeed constitute activism [35] by fertilising seeds of awareness of their own relationships with clothing, and the designer's experiences. Reubens [28: 29–30] highlights that designers have the power to influence the demand of consumers: 'Designers are positioned to shape mainstream value systems [40] and design, because they have the skillset to understand people, and influence their values, attitudes and aspirations [39]: they can design with the intention of influencing people to behave sustainably' [21]. The workshop opened up the space to share experiences and knowledge from both perspectives; the designers and the wearers, and in doing so, opportunities to reflect and reconsider our choices were presented. For example, when discussing material choices for garments, the topic of synthetic fabrics and chemical dyes and their impacts on the environment and human skin were brought to focus. This granted the opportunity to present khadi as a more responsible option.

The workshop also addressed a wider perspective of fulfilling fundamental human needs as developed by Max-Neef [23]. He outlines these needs as subsistence, protection, affection, understanding, participation, recreation/leisure, creation, identity, and freedom. In a manifesto addressing fashion, Rissanen declares the satisfaction of these needs as 'the central goal of fashion design' [29]. The workshop aligned with the need for creation in all four existential categories (Being, Having, Doing and Interacting). This included wearers Being intuitive and imaginative. Having and applying their abilities to compose and design when Interacting with the garments that served as protection and provided an expression of identity. It also invited subsistence in being together and interacting as a community within a social environment. The sense of agency encouraged during the workshop demonstrated that wearers felt they could freely express their identities and values, illustrating a foundation of respect and equality between one another. Their participation resulted in a new and alternative experience that developed further understanding of the role of designer and maker, and brought valuable insights on garment relationships (Fig. 8).

Stepping into the role of designer-as-facilitator [7], and away from a more typically solitary role posed itself as a challenge. It was at times difficult to engage in, record and consider the many interactions that unfolded during the workshop. A suggestion made by one of the wearers for the workshop to be hosted by a group of designers as opposed to a single designer could act as a solution to collecting and processing all the valuable information that surfaced. It could also contribute towards improving the

Fig. 8 Manfred Max Neef's fundamental human needs [23]. Illustration by Jensen Roberts. Published with the authorisation of Jensen Roberts

wearers' experience and support replication (scaling-out) and connection (scaling-up) [22: 180]. The high level of involvement and spirit that the wearers contributed required relinquishing control as the designer, yet fuelled inspiration as opposed to suppressing it. The energetic atmosphere and high levels of engagement were optimum for igniting creativity in both the designer and wearers.

6.4 Further Perspectives

The trialled designer-wearer collaboration showed potential to:

- Develop deeper knowledge on experiences, needs and preferences of garments.
- Encourage self-reflection on garments for designers and wearers.
- Transform the wearer 'into an active user and author, thereby creating new opportunities for fashion to be a satisfier of Max-Neef's ideated fundamental human needs of participation and creation' [29].
- Develop an emotional connection between potential wearers and garments early on.
- Create new experiences within a community and bring community cohesion.
- Actively speculate and offer an alternative socio-material system.
- Contribute towards grassroots activism.
- Instigate behavioural change in relation to design, production and consumer culture.
- Bridge and create connections between members of the value chain.

The exploration unfolded many more questions than insights, underlining that there is a lot more that can and should be done to make meaningful and long-lasting change. To name a few; focus groups could be larger and more diverse, each phase of the community workshop could be a workshop in itself, thus extracting greater and richer insights, workshops could bring together stakeholders to understand common challenges, intentions and visions, further awareness on regenerative natural fibres could be brought to light, and the proposed garment economies could enter a cycle of action and reflection [8: 39] over a longer period of time. Tracing the effects of the process could be done by, for example, studying wearers' relationships with the socially designed and produced garments over the period of a couple of years.

7 Impacts of Rebuilding and Regenerating Relationships

The collaborative case study aimed to create visibility between the relationship between the natural world, human labour and clothing and textiles on a small scale between one designer, one company and a group of wearers. The main findings that centering design practice around rebuilding and regenerating relationships within the value chain has had on an individual scale and a wider scale have been dissected and categorised into the impacts on design practice, and on the design for the sustainability movement by and large.

7.1 A Heightened Synergy Between Designer, Design Process and Material

Six garments that traced the insights and reflections from the collaboration with Khadi London and wearers were translated into khadi. The laborious, hand-powered process of cultivating khadi was a huge influence throughout the design and production process. Walker advocates for design decisions such as the use of materials being the foundations on which our design is built upon, rather than an 'optional add-on' [41]. In this case study, building a relationship with the khadi fabric suppliers and producers led to an increased synergy between materiality and the garments (Fig. 9).

This was seen by, for example, the construction of the garments which was very carefully considered and practiced through a series of prototypes and samples. The handwoven nature of khadi in some cases leads to a looser weave structure than fabrics produced by industrial machines or powerlooms, therefore, durable seams can contribute greatly to garment longevity. Thus, utilitarian construction methods such as flat-felled seams seen on denim garments were utilised, so as to embed technical durability within the garments. A point of consideration was also how to utilise as much of the material as possible, which inspired the decision to embed the selvedges into the design of the garments where possible, such as for pocket and

Fig. 9 Design development in consideration of the insights from the community workshop with wearers

sleeve edges, and hems. In future, this consideration can extend to the development of zero-waste or low-waste pattern-making, thereby utilising as much as or all of the material. In contemplation of garment afterlife, coconut buttons were utilised as closures to complement the indigenously-processed cotton fabric, preserve an interconnectedness with the natural world [29], and to maintain the garments' opportunity to return back to soil at the end of life.

As well as the many opportunities the khadi presented in relation to developing a beautiful visual language through the artefacts, there were also some limitations and challenges. The inconsistency of the fabrics widths along with the aim to utilise as much of the fabric as possible resulted in a slower pace of the design and production phase, and contributed to a more conscious and reflexive design practice that at each stage considered the impacts on resource flows, workers, communities and ecosystems [6: 264].

The surface of the khadi also revealed inconsistencies such as handspun slub yarns, mottled dye and small pieces of trapped hairs, contrasting lint and cotton seeds. However, instead of viewing these discoveries as mistakes in material processing or quality checking, the flaws became embedded stories that told of the farmers, spinners, and weavers whose hands it had passed through, infusing the material with individualism and spirit and helping to break down barriers between designer and the 'other' [36: 114–115]: 'These traces of human interaction and involvement are physical reminders that 'inspire a reduction of physical distance between one thing and another thing; between people and things' [17: 67]. They served as both a challenge and a reminder. A challenge of abandoning our learned aesthetic, and embracing the beauty in the complex and untidy patterns of the natural world and human labour of which we ourselves are part [20: 10]; and a reminder of the rich contribution that khadi makes to rural economies, and the preservation of knowledge, tradition and skills (Figs. 10, 11 and 12).

7.2   Towards Conscious Clothing and Textile Futures

The engagement with khadi, khadi suppliers, and wearers opened up opportunities to shift garment design practice from the often fast-paced process spread across nations and geographies [28: 44], to one which centres around SLOC (Slow, Local, Open, Connected) realities [22: 178–179]. Manzini proposes that these realities need not limit practices to one place within one community, rather as influential nodes 'in the larger global network, open to global flows of people, ideas and information' [22: 177–178]. A growing number of designers and makers are rejecting the norms of the fashion industry—including mass production in factories across the globe, seasonal collections, the sole focus on aesthetics, the use of synthetic fibres—and are instead developing practices that centre around slow and conscious methods of design and production. Fletcher defines the method of 'slow' in relation to fashion as one that understands the fashion sector as a 'subsystem of the larger system of economics, society, and planetary ecosystems' [6: 264] and recognises that these factors have

Fig. 10 Experimental artefacts, Vata trousers, Shala vest, and Kośa shirt, developed in khadi cotton. Handcrafted by Ashna Patel. Photograph by Greta Megelaityte

to be part of the fashion debate. These practices that stand on the many foundations of holistic sustainability, such as collaboration, equity, empathy, awareness, responsibility and transparency, alongside the reduction of speed and scale, may well be shaping a slow and steady redefining of luxury. In this reality, luxury is built on intimate relationships that revolve around ethical community initiatives, a nurturing of

Fig. 11 Experimental artefact developed in khadi cotton. Handcrafted by Ashna Patel. Photograph by Greta Megelaityte

resources, reassociating with nature, and bespoke experiences that engage the mind, body and soul [30].

In parallel, consumers are growing more and more aware of the consequences of the mainstream fashion and textile industry through the increasingly placed importance of and access to research, discussions and exposure of the exploitation and

Fig. 12 Experimental artefacts, Apāna trousers and Kośa shirt, developed in khadi cotton, Handcrafted by Ashna Patel. Photograph by Greta Megelaityte

extraction of workers, environments and communities globally. In the year 2020, where a global pandemic swept across the world and forced everyday life to halt, many retailers and brands collectively refused to pay the 70 million workers that had already produced and shipped billions of dollars worth of clothing. This decision left workers in desperately vulnerable situations, in some cases on the brink of homelessness and starvation, and led to the 'PayUp' movement–a movement which aims for equity and justice within the industry by demanding retailers and brands to pay fair wages to workers, provide transparency through annual data reports and public commitments, sign enforceable contracts and help pass laws to ensure the stability of livelihoods for workers [5]. The injustices of the industry are a force for the gradual shift from quantity to quality, and from invisibility to transparency. A desire for clothing and textiles to be products of all four tenets of sustainability including ecological sustainability, social sustainability, economic sustainability and cultural sustainability [28: 52], and be rooted in awareness, compassion and inter-relatedness, is bringing changes to the ways in which the industry operates. These changes offer handmade materials like khadi, and slowly crafted clothing a place in an industry that is demanding a conscious awakening.

8 Conclusion

Granted, this case study does not serve as a concrete solution or a fixed reality. It is an investigation that begins with the intention to shift relationships and experiences—as members of the value chain—and to engage with materials, processes and the speed and scale of design, production and consumption culture [25]. As Rissanen puts it: 'Possibility is a high potent word: within resides the power to create new realms' [29]. Stepping into unfamiliar territory presented many possibilities. Firstly, the possibilities of actively living into a meaningful collaborative network with those that have the ability and responsibility to regenerate the garment and textile industry. In working alongside those who approach the industry from a values-based stance, the imperative to also be led by this approach was heightened. In the case of this research, working with a natural fibres activist and supplier encouraged and developed the importance of a conscious synergy between materiality and clothing. Through actively learning about the material's history, story, and processing, the knowledge was translated into a deeper connection when engaging and exercising the fabric and a firm under-standing that the materials did not harm its producers, and will not harm its wearers. The collaboration produced information that enabled a more meaningful design prac-tice and birthed the inclination to pass this understanding onto wearers, to inspire their own conscious engagement with the cloth; to value and treat their clothing as long-term companions. Ultimately, the collaboration instigated the narrative of the raw material permeating the garments' narratives, a reassociation between textiles and nature, and an innate understanding that clothing is a product of land and labour.

The grassroots collaboration with wearers shifted the process of cultivating garments into a social endeavour, with the belief that if clothes are meant for people, they should be part of the conversation and cultivation. This required shifting into a new designer role beyond the confines of a design studio. Rebuilding these rela-tionships brought feelings of discomfort and uncertainty. However, creating a space where ideas, experiences and dialogues were shared showed potential to contribute to cultivating garments–and experiences–that fulfil our fundamental human needs, while preserving tradition, supporting rural livelihoods and regenerating agricul-tural land inherited by ancestors. The workshop brought community connectivity, creativity and grassroots activism to name a few. Inviting in a diversity of cultures opened up (rare) dialogues where a variety of perspectives and insights were exchanged. These valuable insights showed potential to enhance the possibility of cultivating emotionally, technically and aesthetically durable garments that may bear increased longevity. The current disconnection between the value chain has correlated in a disconnection between our actions, as practitioners and wearers, and their impacts on both the human and non-human ecosystems. In this regard, efforts to build a meaningful collaborative network hold the potential to deepen the narratives embedded within a garment and may activate new realities that ripple out balancing the ecological, social, economic and cultural systems that bring garments to fruition.

Acknowledgments I would like to extend my deepest gratitude to my supervisor, Professor Ulla Ræbild (Design School Kolding, Denmark) and my mentor, Kishore Shah. This paper and the research behind it would not have been possible without their exceptional knowledge, wisdom, guidance and support. I am also hugely grateful for the support and insights offered by the editors, Dr. Miguel Ángel Gardetti and Dr. Subramanian Senthilkannan Muthu, and the anonymous peer reviewers at Springer Nature.

References

1. Anderson K (2018) 'A frayed history'—a review. https://khadi.london/a-frayed-history-a-rev iew/. Accessed 2 Dec 2020
2. Burgess R, White C (2019) Fibershed: growing a movement of farmers, fashion activists, and makers for a new textile economy. Chelsea Green, Vermont
3. Burgess R, Creque J, Dayaneni G, Bellos S, Chopra N. Defining regeneration for fiber & textile systems. https://fibershed.org/. Accessed 29 April 2020
4. Chapman J (2009) Emotionally durable design. Earthscan, London
5. Dandeniya A, Akter N, Barenblat A, Cline EL (2013) PayUp Fashion. https://payupfashion. com/#actnow. Accessed 20 Nov 2020
6. Fletcher K (2010) Slow fashion: an invitation for systems change. Fash Pract 2(2):259–266. https://doi.org/10.2752/175693810X12774625387594
7. Fletcher K, Grose L (2012) Fashion and sustainability: design for change. Laurence King, London, pp 157–179
8. Fletcher K, Tham M (2019) Earth logic: fashion action research plan. JJ Charitable Trust, London
9. Futane K (2018) Cotton: towards a holistic solution for change. https://khadi.london/cotton-towards-a-holistic-solution-for-change/. Accessed 10 Dec 2020
10. Gandhi MK (1927) The birth of khadi. In: Narayan S (ed) An autobiography: the story of my experiments with truth. Navajivan Publishing House, Ahmedabad, pp 539–556
11. Gaver W, Boucher A, Pennington S, Walker B (2004) Cultural probes and the value of uncertainty. Interact Funol 11(5):53–56
12. Holroyd AT (2018) Forging new futures: cultural significance, revitalisation and authenticity. In: Walker S, Evans M, Cassidy T et al (eds) Design roots: culturally significant designs, products, and practices, 1st edn. Bloomsbury, London, pp 25–38
13. Khadi London (2020) https://khadi.london/. Accessed 04 Dec 2020
14. Khadi Village and Industries Commission (2019) Annual progress report (2018–2019). http://www.kvic.gov.in/kvicres/apr.php. Accessed 15 Dec 2020
15. Khalid A, Aguilar AR, Pérez CM et al (2016) From banks and tanks to cooperation and caring. Movement Generation, Oakland

16. Klepp IG, Bjerck M (2014) A methodological approach to the materiality of clothing: wardrobe studies. Int J Soc Res Methodol 17(4):373–386. https://doi.org/10.1080/13645579.2012.737148
17. Koren L (2008) Wabi-sabi for artists, designers, poets & philosophers. Imperfect Publishing, Point Reyes
18. Kothari A, Venkataswamy D, Laheru G et al. (2019) Sandhani: weaving transformations in kachchh. Kalpavriksh. https://kalpavriksh.org/publication/sandhani-weaving-transformations-in-kachchh-india-key-findings-and-analysis/. Accessed 15 Dec 2020
19. Krogh P, Koskinen I (2020) Drifting by intention. Springer, Cham, pp 50–53
20. Lappé FM (2009) Introduction. In: Fukuoka M, Korn L (eds) The one-straw revolution: an introduction to natural farming. The New York Review of Books, New York, pp 7–11
21. Lockton D (2013) Design with intent: a design pattern toolkit for environmental and social behaviour change. PhD Thesis, Brunel University, London
22. Manzini E (2015) Making things replicable and connected. Design, when everybody designs: an introduction to design for social innovation. MIT Press, Cambridge; London, pp 177–187
23. Max-Neef M (1992) Development and human needs. In: Etkins P, Max-Neef M (eds) Real life economics: understanding wealth creation. Routledge, London, pp 197–213
24. Menon MU (2017) A frayed history. The journey of cotton in India, 1st edn. Oxford University Press, New Delhi
25. Niranjana S, Vinayan S (2001) Report on growth and prospects of the handloom industry. Dastkar Andhra, Secunderabad, pp 1–6
26. Panchal JA, Ranjan MP (1993) Feasibility report on the proposed institute of crafts. National Institute of Design, Ahmedabad, pp 13–22
27. Redström J (2011) Some notes on programme/experiment dialectics. In: Nordes. The Interactive Institute, Helsinki, pp 29–31
28. Reubens R (2019) Holistic sustainability through craft-design collaboration, 1st edn. Routledge, London, pp 7 64
29. Rissanen T (2018) Possibility in fashion design education—a manifesto. Utop Stud 28:528–546. https://doi.org/10.5325/utopianstudies.28.3.0528
30. Sanderson C (2017) The five stages of luxury
31. Shah K (2020) Conversations with khadi london founder, kishore shah
32. Shah P, Amin SS (2017) Cotton and hand crafted fabric processing at village level. In: Revival of khadi summit. Prashant Industries, Ahmedabad
33. Sinift A (2018) Weaving Stories: artists in collaboration with gandhi ashrams (Part 1). https://khadi.london/weaving-stories-inspired-by-jhola-bags-part-1/ Accessed 10 Dec 2020
34. Srivastava A, Khan P, Rawat D, Sri RV (2020) India: khadi reform and development program. The Asian Development Bank, Manila
35. Tham M, Jones H (2008) Metadesign tools: designing the seeds for shared processes of change. In: Changing the change: design, visions, proposals and tools. Goldsmiths Research Online, London, pp 10–12
36. Thomas S (2011) Spirituality and ethics: theopraxy in the future of sustainability within the supply chain. In: Fletcher K, Tham M (eds) Routledge Handbook of Sustainability and Fashion, 1st edn. Routledge, London, pp 111–120
37. Traditional craftsmanship. In: ich.unesco.org. https://ich.unesco.org/en/traditional-craftsman ship-00057. Accessed 13 Apr 2020
38. Trivedi L (2007) Introduction. Clothing gandhi's nation. Indiana University Press, Bloomington, pp 17–26
39. Vezzoli CA, Manzini E (2008) Design for environmental sustainability. Springer Science & Business Media, London
40. Wahl DC, Baxter S (2008) The designer's role in facilitating sustainable solutions. Des Issues 24(2):72–83
41. Walker S (2011) Meaning in the Mundane: Aesthetics, Technology and Spiritual Values. The Spirit of Design: Objects, Environment and Meaning, 1st edn. Routledge, London, pp 163–185

42. Walker S, Evans M, Mullagh L (2019) Meaningful practices: The contemporary relevance of traditional making for sustainable material futures. Craft Res 10:183–210. https://doi.org/10.1386/crre_00002_1
43. Woodward S (2019) Material methods: researching and thinking with things. Sage Publications Ltd., London

A Sustainable Model: Handloom and Community in Meghalaya, Northeast India

Anna-Louise Meynell

Abstract In traditional weaving societies across the world, handloom is so embedded into community life that boundaries are often blurred between livelihood, culture and community. The integrated nature of handloom weaving in agricultural communities of Meghalaya (Northeast India) is the focus of this paper, where the balance of the individual and the community has been maintained over generations, where artisans maintain their individual practice alongside their communal agricultural responsibilities. It argues that this balance is crucial to environmental, social and cultural sustainability in the localised context. While the common narrative of handloom development in India is centred on productivity and livelihood generation, this paper explores the value of a part-time practice, aligned with the changing seasons and responsibilities of agriculture. It draws on the example of the *eri* silk weaving communities of Meghalaya, although many of the observations are common with other communities across Northeast India and beyond.

Keywords Handloom · Textiles · Eri silk · Agriculture · Sustainability · Meghalaya · Khasi · Natural dyes

1 Introduction

The overarching argument of this paper is that the narrative of handloom development initiatives that encourage handweavers towards a more mechanised practice and production-based model requires deeper consideration. The shift from handweaving as a cultural activity to a primary livelihood has proved to be successful in many handloom communities across the world, such as Living Blue in Bangladesh, or Women Weave, or Avani in India, organisations that consider the environmental and cultural impact of their work as well as the economic. However, this paper maintains that without such a holistic approach there are environmental, social and cultural

A.-L. Meynell (✉)
Centre for Sustainable Fashion, London College of Fashion, University of the Arts London, London, United Kingdom
e-mail: info@annaloom.com

© The Author(s), under exclusive license to Springer Nature Singapore Pte Ltd. 2021   165
M. Á Gardetti and S. S. Muthu (eds.), *Handloom Sustainability and Culture*,
Sustainable Textiles: Production, Processing, Manufacturing & Chemistry,
https://doi.org/10.1007/978-981-16-5967-6_8

consequences of handloom development that can go unnoticed. This perspective does not intend to undermine the value of handloom as a primary livelihood, rather through the case study, it celebrates the many artisans in the Ri Bhoi District of Meghalaya who are content with their handloom practice and life balance as it is. It argues for a more nuanced understanding within handloom development, that values handloom not only related to income generation, but as a cultural activity that considers the artisans within the community. It is a perspective that views environmental, social and cultural sustainability to be interconnected on all levels.[1]

1.1   Organisation of the Paper

The paper starts with an introduction to handloom, as integrated within the agricultural community, defining the fields of sustainable development and handloom development in India. It then introduces the case study of *eri* silk weaving in Meghalaya, where a detailed description of the production process highlights the depth of environmental, cultural and social sustainability within the practice. A brief synthesis of development initiatives in the Ri Bhoi District follows, with analysis of the cultural shift that takes place as a more livelihood-based approach to handweaving becomes commonplace.

1.2   Contextualising the Terms

This paper refers to handloom in Meghalaya as being integrated within the agricultural community. Life in rural communities is closely tied to the land, members are born into the community, bound through a shared experience of the land, with a common history, culture and values [10]. Life in the villages of the Ri Bhoi District in Meghalaya follows the cycle of the rice paddy, agricultural responsibilities are shared amongst everyone and many secondary livelihoods exist to support or complement agriculture, such as craft and handloom. Productivity and local economies of rural communities are embedded within this integrated system, connecting the various skillsets, needs and requirements of the people of the community. Economic historian, [20] explains the traditional relationship between farmers and weavers in pre-colonial India:

> "Weaving (....) and the trade and production of cotton cloth cannot be studied in isolation from agriculture (...) The cotton itself was a product of agriculture and much of the textile manufacturing work – from the cleaning and preparation of the cotton to the spinning of the yarn–was done by agriculturalists."

---

[1] The research into the handweaving communities of the Ri Bhoi District of Meghalaya formed the backbone of the author's doctoral thesis, who has been working, researching and interacting with weavers of the Ri Bhoi District since 2014.

Parthasarathi [20]: 6).

Colonial crafts revivalist, Coomaraswamy explored the relationship between craft and agriculture, positioning the rural craftsman as a supplier engaged in a "perpetual contract" with the agricultural community as carpenter, blacksmith, potter and washer man [8: 1]. Handloom in the Ri Bhoi District stands as an example of a supplementary livelihood not so much "in contract" with agriculture, but interconnected at each and every step with farming. Coomaraswamy's definition is significant here because it points to an informal structure of trade, labour exchange or sense of communal responsibility, recognising craft in the ecosystem of rural communities.

For indigenous cultures, environmental sustainability is an unconscious way of life, built into the DNA of the people. UNESCO recognises indigenous knowledge as an increasingly important source of knowledge for climate change assessment and adaptation [29]. There is an increasing understanding of the link between sustaining indigenous cultures and livelihoods and sustaining the environment on a local level. Rural livelihoods by and large depend on local raw materials. For artisans, sourcing raw materials themselves from the physical surroundings creates an understanding for the life cycle of natural materials and the need to replenish whatever they use. They have a vested interest in ensuring biodiversity is not only protected but continues to be available for the following generation. Furthermore, and in consideration of craft, the creation of products from local materials and human resources contributes to a sense of cultural and personal pride [25].

The World Commission on Environment and Development (WCED) defines the concept of sustainable development as:

> "a process of change in which the exploitation of resources, the direction of investments, the orientation of technological development, and institutional change are all in harmony and enhance both current and future potential to meet human needs and aspirations". (WCED [26]: 15)

Sustainable development ensures that development of society takes place with considerate use of the earth's natural resources, that one is not compromised by the other. It acknowledges the need for social and economic development, yet not at the cost of the planet. In 1987, the Brundtland Commission (then known as WCED) defined environmental, social and economic development as the three pillars of sustainable development. These pillars are interconnected, for example addressing climate change through supporting rural livelihoods can divert the trend of urban migration and the strain on the environment which results from a larger urban population. If the economic stability of rural communities is strengthened, many people born and raised in rural areas choose to stay in their familiar environment. Continuity of rural lifestyles ensures social and cultural values are maintained, coming full circle back to the notion of the integrated community's respect for the ecosystem and support of one another.

The Stockholm Intergovernmental Conference on "Cultural Policies on Development" in 1998 elaborated on the importance of culture to human development, maintaining the key message from the World Commission of Culture and Development that "development without culture is growth without souls" (WCCD [27]).

There was a recognition that culture must be considered if development efforts are to be effective [28]. Cultures that intrinsically hold sustainable values provide an internal platform for society to nurture the other three pillars of sustainable development (environmental, social and economic) and to incorporate them into daily life [2].

This paper proposes that craft practices located within agricultural communities nurture sustainable values and offer continuity to cultural understanding and respect of the environment. Further, in relation to handloom development, it should be acknowledged that craft activities never take place in isolation, that an entire "craft culture" exists around it. Walker et al. [25] defines this as the "craft ecosystem" that supports craft practice. [17] refers to the "socio-technical ensemble" that includes all stakeholders of craft production. It recognises the complete process of the craft, beginning with agriculture, ending with marketing and sales of cloth, encompassing each and every step of textile production in between, including development workers and entrepreneurs. The common factor of these terms is the recognition that craft is embedded within a broad social and cultural community, interconnected with other livelihoods and practices. Sustainable development of handloom therefore considers the practice within the entire ecosystem, with the understanding that handloom is more than the practice or the product, or the economics of the livelihood.

In the Indian colonial period, there was an institutional recognition of craft beyond the crafted product, beyond economics and production, that considered the social significance of the craft culture. Craft and handloom became cultural symbols in the context of nationalism and the affirmation of heritage [6, 19, 21, 23]. It came to represent not just a set of products or a type of production, but a larger snapshot of Indian society itself, representing its visual culture, social organisation, intellectual traditions and engagement with the larger world [19]. Craft remains important in India today through economics of employment and national identity, although the value it is given by government in the contemporary development of the nation has shifted from the Gandhian elevation of the craftsman and reverence for their skill, to the adoption of a modernisation narrative, where handloom schemes are focussed on upgrading technology, creating employment and generating income. Recognising this shift lays the ground for the argument that government-driven handloom development has a primarily economic development agenda, as opposed to the holistic approach of sustainable development that considers the environment and culture as well.

1.3 Situating the Case Study: The Culture of the Ri Bhoi District

Meghalaya, meaning the "abode of the clouds" in Sanskrit, lies on the eastern border of Bangladesh (Fig. 1). The Ri Bhoi District is one of eleven districts of Meghalaya, located along the border of Assam. With a lower altitude than the other states of Meghalaya, the Ri Bhoi District gets the best of the climate from both states; the

NORTH EASTERN INDIA

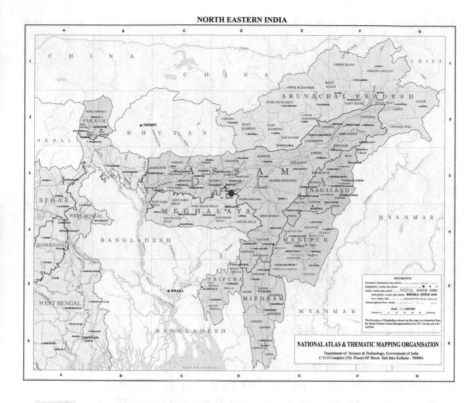

State of Meghalaya indicated in yellow

● Location of the Ri Bhoi District

Fig. 1 Northeast region, Meghalaya & The Ri Bhoi district. (Map source: National atlas and thematic mapping organisation, Kolkata.)

rain from Meghalaya yet without the chill from the highlands, along with the sun of Assam yet without the stifling humidity of the Assamese flood plains. As a result, it is the most fertile of the state of Meghalaya, it is the agricultural heart of Meghalaya (Fig. 2).

The primary ethnic groups indigenous to Meghalaya are the *Khasis*, the *Jaintias* and the *Garos*. A unifying thread amongst them is the matrilineal structure of their societies. Not to be confused with a matriarchal society where women have authority, the matrilineal society reveres the maternal ancestress of the clan. The family blood line is passed on through the woman, and the financial inheritance and tangible assets traditionally go to the youngest daughter of the family. Authority and decision making however, rest with the male members of the clan. The clan, or *kur,* is perhaps the most important social structure in *Khasi* communities, reinforcing the kinship of the wider family and community [22]. The *Khasi Bhoi*, a subtribe of the *Khasi* family is the ethnic community studied in this research. Although widespread conversions

Fig. 2 Rice cultivation in the Ri Bhoi District (Image source: Authors own, Ri Bhoi District, Meghalaya 2016)

to Christianity in colonial times changed the spiritual landscape of Meghalaya, the indigenous *Khasi* religion, *ka Niam Khasi* is practiced by a small minority. It is based on animism and ritual, a belief system closely related to spirits and nature. *Khasi* sociologist, Mawrie describes: "They see a valuable connection between nature and the life of a man. Nature is like a great book with infinite leaves in it. Each page contains valuable lessons which a *Khasi* takes in with relish" [18: 98]. This respect for the environment is deeply rooted in the *Khasi* consciousness regardless of religious allegiance, instilled through ancient *Khasi* mythology still recounted today. *Eri* silk weaving embodies this respect for the environment reflecting the values of the *Khasi* culture, as a physical manifestation of the artisan in sync with her natural environment.

2 Eri Silk Weaving in Meghalaya

Eri silk is indigenous to Northeast India and has been cultivated for centuries by artisans in the region. The matt hand-spun texture is soft and slubby, with an appearance closer to a hand-spun wool or cotton than a shiny sumptuous silk (Figs. 3 and 4). It is a short staple fibre, meaning the cocoon is made up of many short, soft and fluffy strands of silk, rather than the long continuous fibre of mulberry silk cocoon. The characteristic "silky" shine of mulberry silk is due to the long unbroken filament. The

Fig. 3 Eri silk fibre and thread (Image source: Carol Cassidy, Ri Bhoi District, Meghalaya 2014)

only way to loosen the gum that binds the mulberry silk cocoon is to boil it, with the silkworm still inside the cocoon. In contrast, after the *eri* silkworm has completed the spinning process it is either taken out by the artisan tearing open the cocoon, or left to complete metamorphosis, emerging as a moth and continuing the life cycle. As *eri* silk is hand-spun it is of no matter that the cocoon has been broken open, thus saving the worm from being boiled alive. This non-violent production process has led to the widespread marketing of *eri* silk as *peace silk* or *ahimsa silk*. The concept of *ahimsa* is rooted in Jainism, a religious belief system and a way of life that advocates non-violence or disruption of the natural life cycle of any living being, or creature.

2.1 Environmental Sustainability

The production process of *eri* silk, from cultivation of the silk to the finished product demonstrates the ways in which human lives of the handloom communities (in the Ri Bhoi District) are interconnected through processes of production with the lives of animals and plants, weather and the land [15]. It is through the integrated nature of artisanal and agricultural tradition, and knowledge, that *eri* silk emerges as a holistic example of environmentally sustainable textiles, where culture and community are the bedrock of the practice.

Fig. 4 Naturally dyed eri silk cloth (Image source: Carol Cassidy, Ri Bhoi District, Meghalaya 2014)

2.1.1 The Craft Process

The *eri* silkworm is scientifically known as *Philosamia Ricini* or *Samia Cynthia Ricini*. It is a multivoltine species, meaning they can be reared multiple times a year. The main source of food for the silkworm is the leaves of the castor plant (L. *Ricinus Coimmunis)*, essentially a weed that grows in abundance, which needs no irrigation and spreads fast. Cultivation of food plants is rarely organised and leaves are generally collected by villagers from their fields or nearby forests (Fig. 5). In villages where there is high demand for castor leaves for *eri*culture, villagers will take turns cultivating worms, in a manner of rationing the leaves, acknowledging the limits to the resources. As the worms grow, they produce a considerable amount of excrement. These droppings, once dried and crumbled, are a highly effective pesticide on the rice paddy and other crops. This practice, common also to other sericulture communities, evolved over the generations, through observation and experience, through the co-existence of sericulture and agriculture.[2]

---

[2] The agricultural value of the silkworm faeces has been acknowledged by the research community in China, exploring the potential as an organic pesticide (Chen 2011).

Fig. 5 Collecting castor leaves for ericulture (Image source: Authors own, Ri Bhoi District, Meghalaya 2014)

After approximately one month, the worms begin to spin their cocoons (Fig. 6). It takes about four to five days for the cocoon to be spun, and a further five to seven days for the moth to emerge if allowed to complete the cycle of metamorphosis. According to artisans, approximately 20% of all cocoons produced in the Ri Bhoi District are left untouched for the moths to emerge, mate and lay eggs for the next cycle. The remaining 80% are generally removed from the cocoon by tearing the silk and gently pulling the worm out (Fig. 7). These worms will either be eaten by the family or sold in the market as a source of additional income/protein. This is common practice amongst *eri* silk communities of the Ri Bhoi district—it is part of the process, part of the tradition. It does however raise an uncomfortable contradiction with the term

Fig. 6 Eri silkworms spinning their cocoons (Image source: Authors own, Ri Bhoi District, Meghalaya 2015)

Fig. 7 Extracting the worms from the cocoon (Image source: Champak Deka, Ri Bhoi District, Meghalaya 2020)

Fig. 8 Handspinning with the takli (Image source: Iba Mallai, Ri Bhoi District, Meghalaya, 2019)

ahimsa silk since it is most definitely not a peaceful and natural end to the worm's life. Viewed from an alternative perspective, it demonstrates a resourceful zero-waste lifestyle.

Handspinning is done using the drop spindle (*takli*), a small stick with a hooked end and a weighted bottom (Fig. 8). The mastery of handspinning depends on the co-ordination of the spin, the tension and the continuous teasing out of the fibres. Drazin and Kuchler (2015) advocate the recognition of how raw materials are transformed: as fibre becomes yarn, as wood becomes object, as clay becomes bricks; as substance becomes form. They argue that the agency of the materials, the place and the artisan are equally important in the process. This transformation from fibre to yarn is the crucial, initial stage of production that will affect the quality of the final woven cloth. The finer the yarn, the softer the cloth; a regular and controlled twist will enable a smooth weaving process; the regularity of the count of the yarn (accepting the slubs of hand-spun yarn) will create an even density of cloth.

The traditional colours of textiles in Meghalaya are deep yellow of turmeric root, pinkish-red of the resin from the lac insect, and black of iron ore found in the ground in the forests of the Ri Bhoi District and hills of Cherrapungee. Stick lac is a resin that is secreted by the lac insect, as it colonises the host trees. It is then pounded, crushed and mixed with water to extract the dyeing agent (Figs. 9 and 10). The process of dyeing with stick lac produces a hardened resin from the waste lac substance. This is moulded into lengths and used as a resin to repair tools, or to seal packages. Waste from the dye process will go to the compost.

Artisans in Meghalaya have a tradition of using natural plant-based mordants to fix the colour of their dyes. In a market increasingly aware of sustainable practice, these natural mordants offer an environmentally friendly alternative to metallic mordants such as copper, cobalt, chromium and lead which are damaging to the environment [3]. These mordants include the leaf of the Burmese Grape tree (L. *Baccaurea Ramiflora,* known locally as *sla sohkhu*) and the leaf of the *Terminalia chebula* (known locally as *sla sohtung).* Both leaves contain natural tannic acid that fixes the colour, essentially colourless but it can be used to deepen or brighten colours. The bark of a tree known locally as *diengrnong* (L. *Berbis wallichiana*) is a natural mordant that produces a rich olive-green colour when used with the bark of the *waitlum pyrthat* tree (L. *Oroxicum indicum*). When *diengrnong* is used with turmeric dye, the quality of the yellow is bright and luminous. Artisans are both proud and protective of their traditional knowledge of natural mordants since other artisans in the region do not have the same practice. It is common to see saplings of these trees around the homes of the artisans, acutely aware of the need to conserve these precious resources.

The constant give and take between agriculture and weaving can be further observed in the setting-up of the loom, when rice paste is applied to the warp as a strengthening agent to avoid thread breakages during weaving (Fig. 11). The starchy paste is made by boiling rice and squeezing the liquid through a cloth. Leftover rice kernels from the mixture are fed to the pig. The traditional method of weaving *eri* in Meghalaya is on the floor loom, *(thain madan).* The floor loom (Figs. 12 and 13) is a modest structure made of local bamboo and wood. It is made either by the weaver themselves or another community member. The warp is stretched between two sets of wooden posts pounded into the ground and tensioned by a bamboo or wooden beam at the beginning and end of the warp. The string heddles (*luwi*) are setup with each new warp to create the shed between the warp threads (Fig. 14).

Fig. 9 Pounding the natural dye materials (Image source: Authors own, Ri Bhoi District, Meghalaya 2016)

The simplicity of the structure allows for storing the warp away after each weaving session, if required—the tension is released, the warp rolled up until the next time the weaver continues her work. The reed (*snad*) is handmade with a thin bamboo at the top and bottom, and the teeth, or the comb of the reed, are thin strips of bamboo, held in place by tightly wound cotton (Fig. 15). These are all made locally, but not

Fig. 10 Dyeing with the resin of stick lac (Image source: Champak Deka, Ri Bhoi District, Meghalaya 2019)

Fig. 11 Applying rice paste to the warp (Image source: Authors own, Ri Bhoi District, Meghalaya 2016)

Fig. 12 The floor loom (Image source: Authors own, Ri Bhoi District, Meghalaya 2016)

Fig. 13 The floor loom (Image source: Authors own, Ri Bhoi District, Meghalaya 2016)

Fig. 14 Tying the string heddles (Image source: Authors own, Ri Bhoi District, Meghalaya 2016)

Fig. 15 The bamboo reed (Images source: Carol Cassidy, Ri Bhoi District, Meghalaya 2014)

necessarily by the weavers. The fabrication of these fine bamboo reeds brings other members of the community into the wider artisan network using equally developed craft skills to support the tradition of handloom in the Ri Bhoi.

One can attempt to describe, document or understand artisanal knowledge, but putting words to tacit knowledge can never fully capture the depth or the inner sense of an artisan's lived experience. What this research does attempt to illustrate is the sustainable nature of the Ri Bhoi communities, and the manner in which handloom plays an important role in the social, environmental and cultural sustainability. This is a craft culture that relies on nature and the immediate environment for raw materials. Such a direct relationship with the natural world cultivates a practice of replenishing raw materials, of taking care of the forests, fields and all plants. Environmentally sustainable practices are as natural as the breath for these artisans. It is not a conscious effort to live sustainably, it is the lived norm.

2.2   Cultural Sustainability

The various stages of the production process create social links between members of the community, establishing a craft culture which include all individuals who cultivate silkworms and women who spin the yarn, including many who have young children (Fig. 16) or are physically unable to work either in the fields or bent over a loom. It includes people who collect dye plants, natural mordants and firewood, those who craft the tools of the trade and those who may be involved in the sale or promotion of the final product. They may be family members or friends assisting each other, but there are many exchanges of labour, services, product or material that build this network of community livelihoods. The culture of the *Khasi Bhoi* communities is the foundation of these interactions; the centrality of the clan *(kur)* to the *Khasi Bhoi* sense of community creates a strong network amongst artisans and the agricultural mindset facilitates cultivation of the raw materials, with the awareness of the limits of natural resources. The symbiotic relationship between *eri* silk weaving and agriculture has supported the continuation of the craft over generations, shifting energies from craft practice to agriculture with the changing seasons of the agricultural cycle. The planting and harvesting of rice take priority over everything in Ri Bhoi life. These are times when the whole community comes together in the manual labour of the fields, where there is an aligned priority of livelihood activity, where the spirit of community thrives. There is a practice of exchange of labour within the community, where community members take part in each other's cultivation process. During these periods the wider handloom community (spinners, dyers, weavers) puts their practice on hold. This co-existence of agriculture and handloom has emerged as the key factor assisting artisans in the Ri Bhoi District to navigate the challenges of Covid 19 lockdowns, which is discussed later in the paper.

Eri silk handweaving translates this intangible relationship of community and environment into a tangible cloth (Fig. 17). The cloth that is woven, the colours, the designs and the manner of wearing the cloth on the body establishes the visual identity of the *Khasi Bhoi* people, communicating and maintaining their culture. There are

Fig. 16 Young mother weaving on the floor loom (Image source: Authors own, Ri Bhoi District, Meghalaya 2016)

many stripes and checks that are characteristic of the *Khasi Bhoi* (Fig. 18) and supplementary weft designs (Fig. 19) that reflect the interaction with the Assamese tribes settled in the Ri Bhoi District, as distinctly different from the upland *Khasi* textiles. The most distinctive *Khasi* item woven in *eri* silk, common to all *Khasi* subtribes, is the *Khasi* check (*thohrewstem),* the red and yellow gingham check

Fig. 17 Tying the tassels of the *Khasi* check shawl (Image source: Authors own, Ri Bhoi District, Meghalaya 2016)

Fig. 18 Textiles of the *Khasi Bhoi* (Image source: Authors own, Ri Bhoi District, Meghalaya 2017)

Fig. 19 Traditional dress of the *Khasi Bhoi* (Image source: Authors own, Ri Bhoi District, Meghalaya 2017)

shawl (Fig. 17). Any woman wearing it will be recognised to be *Khasi,* or at the very least, from Meghalaya. The plain white *tlem* with 3 ribs at the border is commonly used in honouring guests or VIPs, or as a symbol of friendship or gratitude, once again embedding textiles at the heart of their culture. Traditional clothes are worn primarily at festivals, for dance, ritual or celebration. Although simpler styles of clothing have been adopted for daily wear, maintaining the traditional wear is vitally important for the continuity of their culture, building on the perceived understanding of their cultural roots. There is a deep cultural pride in *eri* silk weaving, since each stage from cultivation through to spinning, dyeing and weaving takes place in their own villages and homes.

> "We feel love, pride and happiness because we do everything by ourselves. We rear the silk worm, spin the thread, give colour, and then weave. We are proud that we can do it all by ourselves and we don't want to discontinue or throw it away, we want to continue as long as we live…… We have to respect our own origin or roots."
>
> (personal interview with artisans of Mawlong, March 2017)

2.3   Socio-Economic Sustainability

Economically, handloom generally comes second or third to agriculture, but without it, many families would fall short. For example, the late Kong Loin Nongpoh, a dedicated weaver from Umtnang, raised her children and grandchildren on earnings from her weaving, as her husband passed away at a young age. She revered *eri* silk as a "gift from God" (personal interview 2015). Kong Kram Makri of Pahambir has a similar story, supporting her family of twelve children and numerous grandchildren, single-handedly through her handloom activities, selling to clients in the local community. The personal histories and relationships with handweaving illustrate the valuable economic role of *eri* silk in the community and suggest a certain respect for the woman's ability to support or contribute to the family income. Within the *Khasi* matrilineal communities, although authority is still within the male control, women have a certain freedom to grow and be successful on their own terms.

Historically, artisans in the Ri Bhoi District would weave for themselves or their extended families, making the occasional sale here and there. *Khasi* historian, Bareh noted weaving in the Ri Bhoi was "a tiny household occupation not concerned with commercial priority" [4:188], noting also "the tendency to make it a part-time profession has kept weaving and dyeing in a state of stagnation" [4: 111]. Although Bareh recounts this with a tone of dismay, the flipside is that weaving was not associated with income, and production time was not related to sales price. The more skilled weavers who had a mastery of the supplementary weft technique for creating motifs, would value the investment of time in weaving a more decorative piece of cloth as an investment in a family heirloom or as a masterpiece demonstrating their skill (Fig. 20). Cultural festivals held weaving competitions amongst local weavers where they would find recognition for their skill amongst the community and the leaders.

Fig. 20 Heirloom eri silk textile of complex supplementary weft designs, (Image source: Authors own, Ri Bhoi District, Meghalaya 2016)

This did lead to some of the more enterprising artisans taking handweaving up as a livelihood activity, yet for the majority of artisans, handweaving was an activity from the heart, taken up for the pleasure of it.

3 Shifting the Balance: Handloom and Development in Meghalaya

Since the turn of the twenty-first century, there has been a shift towards handweaving being an income-generating activity. This is largely directed by the Department of Textiles (formerly Department of Sericulture and Weaving) and other government departments through training workshops and capacity building on business and finance management (Baruah [5]). This has in fact resulted in many more women taking up the craft than previously, albeit with different motives to their ancestors. Yet a shift in handloom practice towards a more livelihood-driven approach does not take place in isolation. The balance of activity within the community, the mindset of individuals and supply of raw materials are impacted by this shift. The transition, however small the set-up, is in fact a major shift in the community. This final section explores the impact of this shift through various examples of the common limitations and bottlenecks in production, of development initiatives to upgrade technology and scenarios of interaction between entrepreneurs and artisans on a professional level.

3.1 A Snapshot of Handloom Development Initiatives in the Ri Bhoi District

Handloom development initiatives in the Ri Bhoi District generally speaking fall under the umbrella of improving the process of handweaving to make a more marketable 'product'. The bottlenecks start at the very beginning of the process, in cultivation itself, with demand of raw material outmatching supply. The Department of Sericulture offers regular trainings on cultivation of *eri* silkworms, and distributes eggs of the silkworm larvae to any individual who requests them. *Eri*culture has become a common income-generating activity for many families of the Ri Bhoi District, but the reality is that tradesmen from neighbouring Assam buy the cocoons in bulk for industrial spinning, leaving insufficient silk for the increasing number of local weavers.

In 2012/3, the Department of Textiles distributed spinning machines to artisans across Meghalaya in a Central Silk Board scheme to modernise and mechanise *eri* silk spinning and speed up the spinning process. Many of these spinning machines can be seen lying idle in handloom workspaces, with artisans reluctant to engage with the technology, facing issues of power supply and mechanical failure. Artisans recounted that no training had been given on use and maintenance of these machines. Furthermore, they acknowledged the social value of handspinning saying "spinning on the machine means you have to stay at home. If you spin with the *'takli'* you can take your work anywhere and have social interactions while you work" (personal interview 2017). In the numbers-based approach to increase production and income it is all too easy to overlook these simple social interactions at the core of rural communities (Fig. 21).

Fig. 21 The social act of handspinning (Image source: Authors own, Ri Bhoi District, Meghalaya 2016)

In order to compete with the Assamese *eri* silk producers, international development agency Deutsche Gesellschaft für Internationale Zusammenarbeit (GIZ) observed that the tactile handle of the yarn produced in Meghalaya could improve (GIZ, [9]. A softer yarn that produces a cloth with a softer drape, would naturally increase the production prospects for *eri* silk fabric, having greater commercial appeal. From 2013 to 2015, GIZ worked with the Department of Textiles, bringing in international consultants to refine the process of degumming the *eri* silk cocoon and handspinning with the *takli*. GIZ took a scientific approach to the analysis of the process and established standards of the practical steps in degumming, bringing more accuracy into the process. The results were commendable, and the improvement to the product was undeniable.

Meghalaya is a treasure trove of raw materials from natural fibres to natural dyes. Natural dye trainings, and exposure trips for artisans outside of the state have made a significant impact on the variety of colours artisans are now producing. The increasing global demand for natural dyes is fuelling a revival of the practice amongst artisans. This increase in demand is a mixed blessing of sorts since the current supply of natural dye plants is limited and this means an increase in demand on the plants. Again, when it comes to availability of stick lac, demand now far exceeds supply, since there are many more artisans using it. Meghalaya Basin Development Authority initiated training in Mawlong village in production of stick lac, which has been met with success. More initiatives like this will be needed to meet the growing demand for this unique natural resource.

In a society that is so intrinsically connected to nature, through spiritual beliefs and through generation upon generation of agriculture-based livelihood and lifestyle, the concept of sustainable use and production of raw materials is not new. The training initiatives detailed above build on this relationship with the natural environment, while keeping the process easy to understand. In contrast, while the introduction of the spinning machines sought to increase productivity, it failed to consider the social nature of handspinning, and was not sufficiently backed up with training to guide artisans through the transition to the machines.

In terms of weaving, government training schemes are largely focussed on the frame loom, though some do still include floor loom weaving. The frame loom (Fig. 22) has greater productivity than the floor loom since it is possible to put on a warp of up to 50 + m. It can also achieve greater regularity and a lighter weight of cloth, offering solutions to quality issues and opening up different markets to weavers. However, with each introduction of new equipment the cultural significance of the cloth shifts and a decline in the traditional practices of floor loom weaving is inevitable. [14], who has written extensively on craft practice and mechanisation, maintains the gradual divorcing of the artisans' sensory interaction from their craft ultimately reduces the unique characteristics of a product, encouraging a standardisation and a replicability of product. While the product of the *eri* silk weavers on a frame loom is still very much based around handloom, every step towards a more production-orientated process is a step towards a more homogenised cloth and a different range of products.

Fig. 22 The frame loom (Image source: Authors own, Ri Bhoi District, Meghalaya 2017)

The frame loom, or fly shuttle loom has been introduced to traditional weaving communities all over the world, which has in many cases altered the structure of weaving communities. [11] refers to the frame loom in Madagascar as being impractical in the lives of most weavers. The frame loom requires more space and longer warps than the *akotifahana* ground loom, which translates into a higher investment. The higher increase in investment is also a limitation for weavers in the Ri Bhoi District, so many are not even coming close to the commercial potential. Gajjala et al. [12] draw on the example of weaving communities in Andhra Pradesh to illustrate how the introduction of the frame loom completely shifted the nature of weaving. The frame loom, being taller than the traditional pit loom could not fit into the domestic spaces of weavers, and was as a result moved to handloom co-operatives rather than domestic workshops, shifting the whole balance of the handloom community. She claims that a simple act of rearranging the physical space "ushers in an entirely new perspective that redefines links between the various actors and slowly may push toward a central point of control to perform the tasks efficiently" (2012: 126). In the Ri Bhoi District, the increase in frame loom production has certainly increased productivity and opened up commercial opportunities for weavers, yet over the six years of interaction with these weaving communities, it has been observed that the frame loom is beginning to be more dominant than the traditional floor loom.

3.2   Social Impact of a Shift to Commerce

The key point here is that while increasing handloom productivity or increasing income-generating possibilities can positively impact rural development, there are other issues that arise alongside it. It can lead to a loss in cultural significance, shifting the dynamic and intention of the craft practice. A commercial outlook can also encourage a system of hierarchy to evolve where previously it may not have existed. A hierarchy where the more productive artisans rise to the top, where those who have a grasp on business can assume positions of traders, or control client relationships. This is a situation found in weaving communities across India, where master weavers maintain a tight control on the orders, interactions, production and financial dealings with clients, locking less powerful weavers into contracts with low wages thereby driving down production prices [24, 12, 17]. In Meghalaya, the scenarios within weaving workshops are not so aggressive, yet in some villages there is certainly evidence of the increase in wealth (from an increase in production) tainting the collaborative community outlook. Umden and Diwon, at an hour's drive from the main highway of the Ri Bhoi District is the hub of *eri* silk weaving in Meghalaya, thanks to significant investment and training from several government and NGOs. There is a jealousy of artisans in surrounding villages, since Umden has received all the training, investment and opportunities. One artisan complained: "almost every household in our village spins thread to sell to Umden village, and Umden gets the name and they are rich". Another artisan (in a village that has received no development assistance) captured the essence of the issue: "There is no competition here, we are all struggling together. Competition only comes when commerce comes" (Personal interview, March 2017). Attitudes are bound to shift with an influences of the ever-changing social world, yet this is rarely considered when implementing a handloom development project-the increase of livelihood potential is the overridding goal.

A widespread problem across all sectors of rural livelihood development in Northeast India is the "hand-out mentality", which has evolved over three decades of development schemes. This refers to the expectation that those below the poverty line will receive free staple foods, electricity or financial assistance to construct a home. That is not to say that this assistance is not necessary, but that in fact it has a negative impact on the motivation to work and creates a sense of entitlement to such handouts. The majority of government trainings for artisans provide a daily stipend to artisans, with good intention of compensating them for their potential loss of earnings in attending the workshop. However, many attend a workshop in order to receive the stipend, rather than to receive the training, and do not incorporate their new skills into their practice. This is a highly frustrating experience for many development workers who impart training expecting to make a difference in the practice of artisans.

> "They don't see far enough to understand the value of changing to suit a market they don't interact with, they don't see beyond their immediate world. Ultimately, they are happy in their village lifestyle, and even when presented with an opportunity to uplift their livelihood, the reality is that many do not want to change."
>
> (Development worker, Shillong, 2019)

This frustration is common to many textile entrepreneurs and designers who face missed deadlines, deviation from orders, or complete lack of delivery even after advance payment. The comment below comes from a local textile entrepreneur whose business ethos is structured around maintaining the traditional craft of *eri* silk handweaving and supporting traditional artisans through professional collaboration.

> "At times I have wondered why did we ever intervene at all? If they do not want our help why are we even trying? Actually, in retrospect, there is a lot that we can learn from them about living in the moment."
>
> (Entrepreneur, Shillong, 2020)

Even for fashion designer Iba Mallai, born and raised in Umden, building a fashion business on the craft tradition of her own village was frustrating. Perhaps being from the same community of artisans herself, she quickly recognised the artisans' priorities were different from her own business-driven priorities. Instead of trying to change their working practice, she acknowledged that changing her own practice was the route to overcome this struggle of dynamics:

> "You cannot interact with them in the same way you would in an office or in industry because they do not understand that. You can't be so professional or expect a professional interaction, you have to be patient and kind. They are motivated by kind words and good relationships. I have found the only way to manage my business alongside managing the weavers is to adjust my own production timetable."
>
> (Iba Mallai, Umden, 2020)

The quotes given above can potentially cast the artisans in a negative light—as having a mentality of resistance, or lack of motivation and professional commitment. However they also indicate a sense of stability of the artisan within their own comfort zone.

3.3   Considering the Artisan Perspective

The examples throughout this paper highlight that training artisans to use different tools is not enough, increasing production possibilities is not enough, even increasing the ambitions of the weavers by talking of outside markets is not enough. In line with [24] perspective, a more nuanced understanding of handloom, including the weavers' motivations, lives, lifestyles and choices is required for any initiative, (entrepreneurial or development) to find success. Artisans' motivations are intrinsically related to their natural, physical and social world, making it all the more important to consider the whole craft culture and community before any intervention in handloom.

The fertility of the land in the Ri Bhoi District means that there is an abundance of produce, all year round. People may not have financial resources, but they do not go hungry. Perhaps this has cultivated a simple sense of security, so the concept of financial savings, putting money aside for the future is alien to them. They live day to day, season to season, revolving around the cycle of the paddy. *Khasi* folklorist,

Kharmawphlang uses the example of the link between rice paddy culture and social mentality, observing: "Cash is reckoned with paddy and while it may be borrowed, the borrower may forget the money owed, he would never forget to return paddy" [16: 35].

It is also worth taking some perspective on scenarios of so-called resistance from artisans to adapt to new market challenges, to appreciate the contentment that appears to exist with so many of the Ri Bhoi weavers, to value the fact that they are not driven by any sense of capitalist greed, or simply the concept of "more". Perhaps this is due to the absence of non-essential "things" in their lives, they are content with what they have, and are not driven by the hunger to acquire more. There are few societies in the world that have not been drawn into the concept of "more"—more money, more orders, more opportunities—so a contradictory question regarding development arises: should we really be trying to change this? Or with greater consideration of the localised community scenarios, can handloom development be approached in a more holistic way, that nurtures the part-time weaver rather than repeating the narrative of productivity to address income generation? What is beginning to become apparent is that the productivity approach is not as stable as originally thought, since with the shift to a full-time livelihood, artisans slowly become dependent on this single source of income.

The global crisis of Covid 19 has highlighted the ironic contradiction of this productivity approach and the shift towards weaving as a full-time livelihood, targeting external markets. Due to the globalisation of past decades, marketing structures of all sectors have become organised on a global level, diminishing the resilience of producers [13]. This is true of craft as much as any sector, with craft organisations pursuing enticing foreign markets. The long periods of global lockdown have had a devastating impact on artisans across the globe. In Northeast India, agricultural activities were not included in lockdown restrictions and as such, many part-time artisans who weave only for themselves and their own communities have felt a significantly lesser impact of Covid 19 on their practice. The ebb and flow of their practice is accepted, where they expect to pause weaving during plantation and harvest. The lack of income from handloom during these months is normal as they invest their energies in the income they can derive from the rice paddy. In contrast, full-time artisans who had succeeded in positioning themselves in markets beyond their own state or country have faced unprecedented challenges in the sudden drop-off of orders and sales. It is in such scenarios, when transport and movement are limited, and accessing outside markets is problematic, that local markets re-emerge as a more stable market for handmade products. Kong Tander Tmung, a successful artisan entrepreneur from Umden continued to work on orders from her local clients in Shillong throughout the lockdown, receiving orders through phone calls, WhatsApp and social media, even offering a home delivery service once complete. The benefits of building a strong local client base have never been overlooked by Ri Bhoi artisans. This wisdom should be celebrated.

An All India Artisans and Craftworkers Association (AIACA) report on the impact of Covid 19 on artisans in India concludes that there is an urgent need to reposition sales and marketing strategies, and a need to repurpose products for the changed environment post-Covid (AIACA [1]). In the case of the Ri Bhoi District, there is ample room to strengthen local markets. The Department of Textiles (in co-operation with other government departments) has created a structure for sales through local craft exhibitions, and offer financial support for artisans to participate, yet there is further scope in cultivating the market linkages with local urban clientele. Rural markets could receive more attention from local government to develop networks, championing local product, promoting local markets in rural areas. A greater catchment area may provide incentive for local artisans to develop their product, for urban shoppers to visit rural markets and for tourists to experience the enchanting indigenous markets, coming away with a local handmade product. A monthly textile market set within the Nongpoh market (the key trading market of the Ri Bhoi District) could draw in a wide clientele if well publicised and organised. Such markets allow artisans to sell product as and when they have stock, rather than pushing them into problematic commercial deadlines.

4 Conclusion

This paper has used the case study of *eri* silk handweaving in the Ri Bhoi District of Meghalaya to lay out the complex ground of handloom intervention in indigenous communities. The description of the practice demonstrated the implicit nature of sustainability in the craft, in terms of respect for the environment and maintaining natural resources. Socially, the craft creates a network of community members interacting to support one another, or engaged in trade within the community. The craft of handweaving and *eri* silk textiles reinforce cultural sustainability, embedding the practice at the heart of the Bhoi culture. Above all, the constant give and take between agriculture, agricultural responsibilities and handloom practice has stood the test of time; a sustainable lifestyle that does not put all the eggs in one basket. It is a model that is built upon the relationship of handloom weaving within the community, that represents integrated sustainability of the Ri Bhoi District.

The argument emerging from the second half of the paper is that it is essential for development initiatives in the Ri Bhoi District to consider the artisans within this integrated community, to recognise the community approach towards livelihood and to build on existing skills and passions that strengthen the artisans' relationship to their environment. There is vast scope for increasing cultivation of natural dye resources, aligned with rural practices of cultivation. Handweaving as a livelihood has the potential to make a positive impact in poverty alleviation, yet not at the cost of the local biodiversity, increasing demand on raw materials, or raising expectations of artisans and introducing a money-driven attitude into the community. This research does not devalue the success stories of entrepreneurship and production-orientated projects in the Ri Bhoi District, rather it illustrates that there is room and potential to develop a different framework of development that encourages a part-time approach, that nurtures local markets and that champions local goods to the local clientele. The example of part-time weavers of the Ri Bhoi District celebrates a weaving culture that does not feel the need to chase external markets, of artisans that are content with their lifestyle, their rhythm of work and their sense of balance within their own environment.

References

1. All India Arts and Crafts Association (2020) Impact of Covid 19 on artisans and crafts enterprise: part II. AIACA, New Delhi
2. Baldacchino L, Cutajar C, (2011) The artisan: a sustainable entrepreneur. In: Rizzo S (ed) Green jobs from a small state perspective. Case Studies from Malta, Green European Foundation, Brussels, p 20–32
3. Banerjee AN, Kotnala OP, Maulik SR (2018) Dyeing of eri silk with natural dyes in presence of natural mordants. Indian J Tradit Knowl 17(2):396–399
4. Bareh H, (1991) The art history of Meghalaya. Agam Kala Prakashan, Delhi
5. Baruah S, (n.d.) Umden eri cluster, Meghalaya. http://www.sriparna.in/umden-eri-cluster-meg halaya/ Accessed Jan 2020
6. Bayley CA (1986) The origins of Swadeshi. Cloth and indian society 1700–1930. In Appadurai A (ed) The social life of things; Commodities in social perspective. Cambridge Press, Cambridge
7. Chen X, Xie Y, Luo G, Shi W (2011) Silkworm excrement organic fertiliser: its nutrient properties and application effect. J Appl Ecol 22(7):1803–1809
8. Cooraswamy A (1909) The Indian craftsman. Probsthain & Co., London
9. Costa M (2013) Greening of the silk value chain in Meghalaya and Nagaland. Report for Deutsche Gesellschaft für Internationale Zusammenarbeit (GIZ), Shillong
10. Dhamotharan M (2009) Handbook on integrated community development–seven D approach to community capacity development. Asian Productivity Organization, Tokyo
11. Etheve A (2019) Madagascan looms. In: Feng Z, Sardjono S, Buckley C (eds) A world of looms: weaving technology and textile arts. Zhejiang University Press, Hangzhou, pp 145–147
12. Gajjala R, Niranjana S, Sayamasundari B (2012) Framing the loom: an indian concept. In Gajjala R (ed) Cyberculture and the subaltern: weavings of the virtual and real. Lexington Press, Minneapolis, pp 109–134
13. Ibn-Mohammad T, Mustapha KB, Godsell J, Adamu Z, Babatunde KA, Akintade DD, Acquaye A, Fujii H, Ndiaye MM, Yamoah FA, Koh SCL (2020) A critical analysis of the impacts of COVID-19 on the global economy and ecosystems and opportunities for circular economy strategies. Resour Conserv Recycl 164(2021):105169. https://doi.org/10.1016/j.res conrec.2020.105169 Accessed May 2021
14. Ingold T (2000) The perception of the environment: essays on livelihood, dwelling and skill. Routledge, London
15. Ingold T (2012) Toward an ecology of materials. Annu Rev Anthropol 41:427–442. https:// doi.org/10.1146/annurev-anthro-081309-145920
16. Kharmawphlang D (2016) Essays in Khasi Folkloristics. Ri Khasi Book Agency, Shillong
17. Mamidipudi A (2016) Towards a theory of innovation in handloom weaving in India. Doctoral thesis, Maastricht University
18. Mawrie BL (2014) Khasi ethics. Vendrame Institute Publications, Shillong
19. McGowan A (2009) Crafting the nation in colonial India. Palgrave MacMillan, New York
20. Parthasarathi P (2004) The transition to a colonial economy: weavers, merchants and kings in South India 1720–1800. Cambridge University Press, Cambridge

21. Roy T (2007) Out of tradition, master artisans and economic change in colonial India. J Asian Stud 66(4):693–991
22. Nongkynrih AK (2002) Khasi society of Meghalaya, a sociological understanding. Galaxy Book Centre, Shillong
23. Tarlo E (1996) Clothing matters: dress and identity in India. Hurst & Co, London
24. Venkatesan S (2009) Craft matters. Orient Blackswan Pvt Ltd, New Delhi
25. Walker S, Evans M, Mullagh L (2019) Meaningful practices: the contemporary relevance of traditional making for sustainable material futures. Craft Res 10(2):183–210
26. World Commission on Environment and Development (WCED)1987) Our Common Future. United Nations, Oxford
27. World Commission on Culture and Development (WCCD) (1996) Our creative diversity. Report of the World Commission on Culture and Development. UNESCO, Paris
28. UNESCO (2013) The Hangzhou declaration: placing culture at the heart of sustainable development policies. Hangzhou. http://www.unesco.org/new/en/culture/themes/culture-and-development/hangzhou-congress Accessed May 2021
29. UNESCO (n.d.) Indigenous people—sustainable development and environmental change. https://en.unesco.org/indigenous-peoples/sustainable-development Accessed May 2021

The Influence of Starch Desizing on Thermal Properties of Traditional Fabrics in Anatolia

Meral Isler, Derya Tama Birkocak, and Maria Josè Abreu

Abstract In this book chapter, three traditional fabrics in Anatolia, namely, Feretiko, Ayancik linen and Burumcuk fabric were analysed in terms of thermal comfort properties. The composition of these fabrics was 50% hemp–50% cotton Feretiko, 50% linen–50% cotton Ayancik linen and 100% cotton Burumcuk. In order to prevent abrasion during the production of these fabrics, cotton yarns used as warp yarns were treated with traditional sizing recipes. For all fabrics, in traditional methods, there is no starch desizing process after weaving. In order to determine the effect of starch desizing on thermal properties of these traditional fabrics, the thermal tests were conducted before and after enzymatic desizing process. The test results of thermal conductivity, thermal absorptivity, thermal resistance, air permeability as well as clothing insulation were presented. Regarding the obtained results, the contact angle values of all fabrics were lower after desizing process, as it was expected. With respect to the thermal resistance values, Ayancik linen had the greatest thermal resistance value as well as fabric thickness value after desizing. Feretiko had the greatest air permeability results for both states, before and after desizing, due to its high porosity with the lowest fabric thickness, fabric weight as well as the yarn number.

Keywords Feretiko · Ayancik linen · Burumcuk fabric · Hemp · Traditional fabrics in Anatolia · Wettability · Air permeability · Thermal properties

M. Isler (✉)
Faculty of Architect and Design, Fashion Design Department, Selçuk University, Konya, Turkey
e-mail: meralisler@selcuk.edu.tr

D. Tama Birkocak
Textile Engineering Department, Ege University, Bornova, Izmir, Turkey
e-mail: derya.tama@ege.edu.tr

M. J. Abreu
Textile Engineering Department, University of Minho, Campus de Azurem, Guimarães, Portugal
e-mail: josi@det.uminho.pt

1 Introduction

As long as human beings existed, they felt the need to protect themselves against natural conditions. Fabric weaving emerged from the needs of people such as protection from the cold, dressing up, covering and prinking up. It has progressed according to the welfare level, living conditions, artistic and technical abilities of each nation [37].

The spindle whorls and loom weights obtained from different cultures at various years prove that weaving in Anatolia started from the Neolithic Age and continued with a continuous progress in the following ages [22]. The oldest known examples of fabrics used in Anatolia have been unearthed from the excavations of Çayönü, Çatalhöyük, Kuruçay and Ališar [50]. Up to the present, in the archaeological excavations in Anatolia, the earliest weaving traces were found on the sickle handle shaped from deer horn in Çayönü (6650 - 6350 BC). The found fabric was linen with good weaving, which was flat, thin and plain weaved.

Among the findings uncovered during the researches in Çatalhöyük, pieces of fabric were found that were not determined as wool or mohair. Although most of the pieces had a smooth thin texture, it has been observed that they have a shawl-like texture due to their spaced wefts. In addition, a fringed skirt with copper tubes hanging from the ends was found on a woman buried in a temple. In another temple, a sculpture with a short tasselled skirt was found. Although there is no clear information about the colours of the weavings due to their carbonization, it is thought that they also dyed their fabrics regarding to the wall paints on the temple walls (Fig. 1) [13].

Anatolian textiles have a very important place among world fabrics in terms of weaving, material and pattern richness. In the historical development process of Turkish fabrics, Seljuks, who lived together with the Byzantine Empire since the twelfth century, gave an important acceleration to the art of weaving in Anatolia [44].

Although Anatolia was a well-known fabric production centre in the Middle Ages, there are almost no fabric samples remained from this period. The most important piece known is a velvet woven piece with an inscription on it that it was made for Alaeddin Keykubat. This piece is exhibited today in the Lyon Weaving Museum. It is accepted that this fabric, which has lion motifs woven with gold wire on a red

Fig. 1 The wall painting in Çatalhöyük (left) and the fabric piece (right) [13]

background had been woven on the special weaving looms of the Seljuk palaces [38, 49].

In fourteenth-century fabrics, motifs are large, colours are limited but vivid. On the contrary, in the fifteenth century, the patterns are smaller whereas the colours are more diverse. The most developed period of Ottoman fabric art is the sixteenth century as in other art fields. The seventeenth century is a period of stagnation. In the eighteenth century, western influences started, and in the nineteenth century, European fabrics were assimilated into Anatolia [41].

Today, the production of local fabrics in Anatolia has mostly been replaced by industrial textile production. Some of the local fabrics are still produced by traditional methods and used for commercial purposes, some are produced with individual efforts for special reasons such as artistic purposes, etc., whereas some of them are no longer produced.

The book chapter was organized as follows. The first section reveals the traditional fabrics in Anatolia in terms of raw materials, used handlooms and production techniques. In the following sections, the three selected fabrics were explained in detail and the information about enzymatic desizing process was given. Afterwards, the results of wettability, thermal conductivity, thermal resistance, thermal absorptivity, air permeability and clothing insulation tests were discussed in order to evaluate the thermal properties of those fabrics.

2   The Local Fabrics and Their Productions in Anotalia

In the early ages of weaving, people placed a long horizontal stick on two fork sticks, dangling the warp threads with heavy stones on each end and passed the weft threads diagonally with the help of their hand and found the simple weaving called plain weave today [7].

This primitive weaving technique, whose weaving direction is in vertical position, has developed and transformed into a loom form around 4000 BC. After a while, it was turned to horizontal position and it has been used in this way since 5000 years [11, 34].

Nowadays, in the production of traditional fabrics, manual, semi-automatic and automatic shuttled handlooms are used. The whipped handlooms, which is the most common used type of shuttle handlooms, are mostly preferred in the production of local fabrics. In these handlooms, the shuttle is thrown with a whip, not directly by hand. These types of handlooms are faster and more efficient than other handlooms. Its design ability is wider due to the number of feet it varies from two to eight in these handlooms [7].

In Anatolia, as in every part of the world, the colours and patterns used in the unique traditional weaving of each region are important elements that contain the cultural traces of the societies and contain historical documents about their past. With the settled life, people started to manufacture those fabrics in a room of their houses, courtyards or in suitable areas, which continued the existence of Anatolian weaving tradition [5].

In the traditional fabrics, different materials are used according to the characteristics of the region. The main factors that create the regional characteristics are the climate of the region, its ethnic structure, the traditional lifestyle, the availability of raw materials as well as the types and production techniques. For instance, wool is used in one region and cotton in the other, while in some places linen, silk and hemp are used [3, 35]. Some examples of weaving have kept their original identity unchanged, some have completely changed due to immigration, cultural interactions, etc. and some have disappeared or are about to vanish (Akpınarlı and Başaran, 2012).

Throughout history, the number of fabric types produced and used in Anatolia is quite high. Especially in the Ottoman period, they were divided into two groups as palace fabrics (Fig. 2) and local fabrics. Kemha, Velvet, Çatma, Seraser, Kutni, Atlas, Serenk, Sevayi, Zerbaft, Diba and Hataiare are some examples for palace fabrics whereas Aba, Alaca, Burumcuk, Çit, Kebe and Beledi are the examples for local fabrics [24, 28]. Kemha is a thick fabric; its warp and weft are mostly from silk but sometimes from gold or silver. Çatma is type of velvet; the embellishment is puffier than the ground. Diba is weaved with the silver or gold mixed silk threads. Zerbaft is a golden weave whereas Atlas is bright coloured silk fabric [45].

Today, it is possible to encounter local fabrics in many regions of Anatolia. Local fabrics are named with the regions they are produced such as Şile cloth, Buldan cloth, Rize cloth (Feretiko), Ayancık linen, Ödemiş silk, Nikfer cloth, Kutnu of Antep, Ehram of Erzurum, Kandıra cloth and Tokat cloth.

The local fabrics presently produced can be examined in four groups:

1. Seasonal fabrics are produced by individuals in the light of past knowledge and used for personal consumption.
2. Fabrics are produced and put on the market in public education centres.

Fig. 2 Some fabric samples of Ottoman period [38]

3. The fabrics commercially produced by the people who received training in the courses opened with the aim of creating job opportunities by reviving the weaving culture in the region.
4. Fabrics designed as a product of local projects conducted by institutions, organizations or individual initiatives to create regional cultural souvenirs [2].

In this book chapter, three local fabrics are produced traditionally, namely, Feretiko, Ayancik linen and Burumcuk fabric were selected and analysed. Currently, public education centres give courses to the local people how to produce these fabrics in order to raise new producers and to contribute to the tourism and culture of the region.

2.1 Feretiko (Rize Fabric)

Since ancient times, hemp grown and the yarns obtained from the stem fibres of this plant have an important place in the Black Sea Region in terms of traditional weaving. The homeland of the hemp plant is Central Asia and Khorasan. At 3000 s BC, the cannabis plant was used by the Chinese, later on it was carried to Europe by the Scythians with migrations to the west. With the arrival of the Turks in Anatolia in 1071, the weaving culture continued to develop in these lands [43]. It is also stated in the literature that the fabrics taken from Black Sea Region (especially form Rize City) were used by the sultans in the Ottoman palaces due to their good mechanical and comfort properties [36].

The natural colour of the hemp threads grown in the region is between cream and brown tones. In previous years, the fabrics were produced with the hemp threads without any chemical process, and bleaching was done by placing them into seawater. For this process, the woven fabric was thrown into the sea or the stream, and afterwards was taken out and dried. This bleaching process was repeated for several times until the colour of the cloth turns white. The low levels of ozone in stream water and chlorine in seawater caused the bleaching process to take 20–30 days. During this process, especially sunny weather was chosen for the fabric to dry quickly [53]. Nowadays, Feretiko (Rize fabric) is made of 20 Ne or 24 Ne single-ply cotton yarn as warp, and its weft is made from 20 Ne hemp yarn. It is a traditional plain weave fabric, which has warp density between 18 and 20 in cm, and weft density between 16 and 18 in cm [51] (Fig. 3).

2.2 Ayancik Linen

Ayancik linen is a type of weaved fabric produced in traditional handlooms by the yarns obtained from the linen plant in Sinop City. Ayancik District is famous for its local woven fabric as "Ayancik linen". The natural colour of the linen thread is

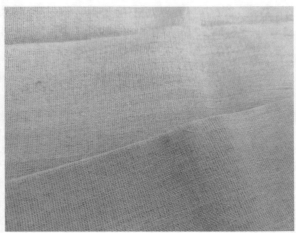

Fig. 3 The handloom for Feretiko production (left) [42] and the fabric sample of Feretiko (right)

brownish, matt and dark. This colour is changed to the cream colour by boiling with wood ash and bleaching. Since flax yarns used in linen weaving are twisted by hand, they are thicker than the cotton yarns used as warp yarns. Three types of weaving are made in the region: linen in both warp and weft, cotton in both warp and weft (Burumcuk fabric), and cotton in warp and linen in weft [16].

The linen yarns are turned into yarn on a spinning wheel and afterwards are put into a large copper cauldron. They are boiled by mixing ash in order to perform bleaching process. After this process, the yarns are wrapped in a wooden mechanism called "elepçek" and then on bobbins. The cotton or linen yarn used in warp is used by sizing [14].

A mixture of flour, soap, starch and oil is prepared with cold water and then boiled with the cotton threads thrown into this mixture. The boiled cotton hanks then are hanged and are left to dry. Afterwards, they are coiled and the warping process is done by adjusting the desired dimensions from the bobbins [1].

In traditional methods, Ayancik linen is woven on handlooms, locally known as "duzen", which are made entirely of wood by interlocking without using nails (Fig. 4). The handlooms are formed from four frames for warp yarns. The comb is made of wood and its teeth are made of reed. Sparse combs are used for peshkir (towel) weaving, and dense combs are used for clothing. The construction of combs and the bobbins used in weft and warp are made of spear. The fabric density differs between 16, 14, 2 yarns per cm^2 in these linen fabrics. Afterwards, the warp threads are placed in the handloom, the shed is adjusted, the weight stone is attached, and the weaving process is performed [48].

Fig. 4 The handloom for Ayancik linen production (left) [48] and the fabric sample of Ayancik linen (right)

2.3 Burumcuk Fabric

Burumcuk fabric is known as shirting fabrics as it is often preferred to be used in shirts production for its good comfort properties. It has been manufactured in different names according to the region such as crepe, meles, helali, heril, satraç, caster, guvul, hand curl and wrinkle cloth [23]. Burumcuk fabric is a narrow width fabric, which are weaved with a width between 45 and 90 cm in handlooms (Fig. 5) [26].

In the past, Burumcuk is used to be produced using silk in warp and weft; however, nowadays it is mostly produced by cotton yarns. In Burumcuk fabric, generally multi-twisted yarns are used as wefts, but also there are examples those yarns are used as both weft and warp yarns. It mostly has between 600 and 800 yarns in the total width. The fabric acquires its wrinkle effect after wetted and rubbed by hand in hot soapy water. The fabric takes its name from the wrinkles that make up its own feature. The width of the fabric shrinks after this process with respect to the twist ratio of the yarns. The air layer which occurred between the garment and the body due to its wrinkle effect provides a better heat insulation [25]. In this study, Burumcuk fabric produced in Sinop Ayancik Region was used.

3 Thermal Properties of Traditional Fabrics

In this chapter, 50% hemp–50% cotton Feretiko, 50% linen–50% cotton Ayancik linen and 100% cotton Burumcuk were analysed. Cotton yarns were used as warp yarns for all fabrics, which were treated with traditional sizing recipes. The sizing

Fig. 5 The handloom for Burumcuk fabric production (left) [48] and the fabric sample of Burumcuk fabric (right)

Table 1 The fabric weights, yarn numbers and the densities of sample

	Fabric weight (g/m²)	Knitting structure	Yarn number (Ne)		Density (yarns/cm)	
			Weft	Warp (cotton)	Weft	Warp (cotton)
Feretiko	105,63	Plain weave	20	60/2	20	20
Ayancik linen	248,73	Plain weave	8	40/2	16	16
Burumcuk	262,96	Plain weave	14	14	22	22

bath was prepared and boiled with starch, flour, oil and soap bar, and then hanks were placed inside for a while. Afterwards, the hanks were hanged and after drying, they are coiled and used in handlooms. For all fabrics, in traditional methods, there is no starch desizing process after weaving. Table 1 shows the fabric weights, the yarn numbers and the fabric densities regarding warp and weft directions.

In order to determine the effect of starch desizing on thermal properties of these traditional fabrics, the thermal tests were conducted before and after enzymatic desizing process.

3.1 Enzymatic Desizing

The fabrics assessed in the present research are produced in handlooms. In order to prevent abrasion in warp yarns and other types of stresses generated during the weaving process, traditional sizing is used with the traditional recipes including starch, flour, oil and soap bar.

It is a fact that the sizing process has effect on the physical properties of the fabrics (Metha 2012) and in this research, its effect on thermal properties was evaluated. The enzymatic desizing process was performed using house-type washing machine (Worten LK-450 T, Portugal), at 70 °C, for 60 min. The solution was prepared with 2 mL/L of enzyme Bactosol at pH 5–6, and Kieralon 1 g/L as wetting agent, and for 1 gr of fabric, 30 mL of solution was added to the desizing bath. After desizing process, the fabrics were rinsed with hot water.

3.2 Mechanical Properties of Traditional Fabrics

Table 2 shows the test results of fabric weight, yarn number and the fabric thickness before desizing (B-D) and after desizing (A-D) process. Sizing is a process applied to the yarns to provide temporary protection and the yarns are strengthened by the sizing bath to reduce yarn breakage during weaving process.

As result of sizing, the yarns become stiffer and this affects fabric construction in terms of being uniform as well as the fabric density. For Feretiko and Ayancik linen, the fabric weight, and so the fabric thickness values are becoming higher after desizing process. However, in contrary, Burumcuk has lower values of the fabric weight and the fabric thickness (Fig. 6).

Burumcuk fabric is woven from double spun cotton yarn on narrow handlooms and afterwards the surface is folded by boiling technique to give the wrinkle effect [23, 37]. Therefore, after desizing process, there remained less wrinkle effect. Thus, during the fabric weight tests after desizing, the samples had less amount of warp and weft yarns despite the fabric density was greater than before desizing. It can be seen in Fig. 7, which presents the difference in fabric view of Burumcuk for before and after desizing process.

Table 2 The fabric weights, yarn numbers and the fabric thicknesses of samples before and after desizing process

| | Fabric weight (gr/m^2) | | Fabric thickness (mm) | | Fabric density (yarns/cm) | | | |
| | | | | | Weft | Warp | Weft | Warp |
	B-D	A-D	B-D	A-D	B-D	A-D	B-D	A-D
Feretiko	105,63	113,37	0,419	0,489	20	20	22	22
Ayancik linen	248,73	274,03	0,703	1,140	16	16	16	18
Burumcuk	262,96	214,5	1,363	1,010	20	20	22	22

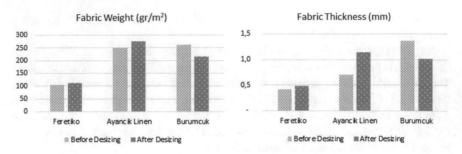

Fig. 6 The fabric weight (left) and the fabric thickness (right) values of the evaluated fabrics

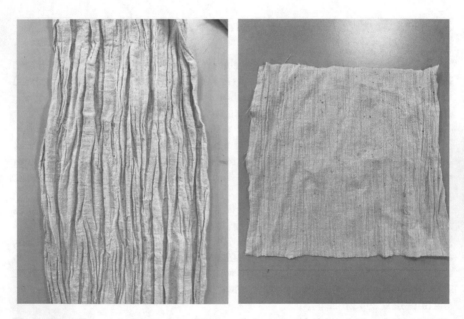

Fig. 7 The difference in fabric view of Burumcuk for before desizing (left) and after desizing (right) process

3.3 Wettability

Wettability has an important role influencing textile's performance; besides different industrial processes such as oil recovery, lubrication, liquid coating, printing and spray quenching [54], it is likewise an important factor in clothing comfort evaluation with the parameters of water absorption and moisture venting [8]. Therefore, in textile industry, the movement of water on textile surface is of increasing interest besides evaluating the hydrophilicity or textiles [19]. Contact angle is the primary value, which indicates the degree of wetting when a solid and liquid interact. Contact angle, shown as θ, is the angle between the solid surface and the tangent to the water surface on the solid. As presented in Fig. 8, θ is defined by the mechanical equilibrium of

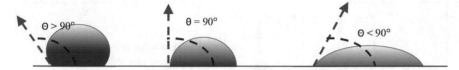

Fig. 8 The contact angle of a liquid droplet on a solid surface (adapted from [33]

the droplet under the action of three interfacial tensions: liquid–vapour (γ_{LV}), solid–vapour (γ_{SV}) and solid–liquid (γ_{SL}) and is calculated by the following equation, where $\theta\gamma$ represents the equilibrium contact angle [39, 54]:

$$\gamma_{LV}\cos\theta_y = \gamma_{SV} - \gamma_{SL} \tag{1}$$

Small contact angles, if γ_{SV} is larger than γ_{SL} (less than 90°), correspond to high wettability means the fluid will spread over a large area of the surface and the water will wick into the surface by capillary action. Large contact angles, if γ_{SV} is smaller than γ_{SL} (greater than 90°), correspond to low wettability that the water will not rise by capillary action and the fluid will form a compact liquid droplet [39]. For superhydrophobic surfaces, water contact angles are usually greater than 150°, showing almost no contact between the liquid drop and the surface [54].

The effect of starch desizing on wettability of fabrics was tested using contact angle measuring instrument (DataPhysics Instruments, Germany) with dedicated software allowing the calculation of contact angles.

It is a fact that sizing process affects the hydrophilicity of the yarns and makes the fabrics more hydrophobic. Therefore, as it was expected, the contact angle values were lower for all fabrics after desizing process (Table 3).

The highest difference was observed for Ayancik linen, almost no water drop was observed on its surface (Fig. 9). Not only the material but also the flatness of the textile surface has an important effect on contact angle values. In the novel method developed by [40], hysteresis was obtained between advancing and receding contact angles which are caused by the surface roughness and/or heterogeneity of the natural fibres. They stated that the contact angle hysteresis for fibres with a rough surface varied from 22 to 30°. This might be the reason having lowest contact angle values difference for Burumcuk fabric obtained before and after desizing process. As it was presented in Fig. 7, after desizing process the surface heterogeneity of Burumcuk changed.

Table 3 Contact angle values of evaluated fabrics	Contact angle (degree)	
	B-D	A-D
Feretiko	121,76	71,01
Ayancik linen	112,76	35,00
Burumcuk	107,81	84,30

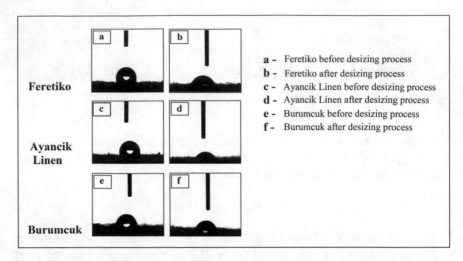

Fig. 9 Contact angles of all samples obtained by contact angle measuring instrument

In Fig. 9, it can be clearly seen that all the samples showed low wettability properties before desizing process due to the contact angle values greater than 90°. When enzymatic desizing process was performed, all the samples had the contact angle values less than 90°, which corresponds to high wettability.

3.4   Thermal Conductivity

The test results of thermal conductivity, thermal absorptivity, thermal resistance as well as the air permeability were presented in Table 4. The thermal properties of fabrics evaluated via an Alambeta tester (Sensora Instruments & Consulting, Czech

Table 4 The thermal conductivity, thermal absorptivity, thermal resistance and air permeability values of evaluated fabrics obtained before and after desizing process

	Thermal conductivity (W/m/K)		Thermal absorptivity (W s$^{1/2}$/m^2 K)		Thermal resistance (m^2 K/W)		Air permeability (l/m^2 sec)	
	B-D	A-D	B-D	A-D	B-D	A-D	B-D	A-D
Feretiko	34,529	40,029	95,27	109,243	0,120	0,122	2686	1565
Ayancik linen	41,171	46,789	110,51	108,211	0,170	0,274	1361	795
Burumcuk	42,529	44,375	87,89	105,2	0,321	0,245	559	482

Republic), which simulates the dry human skin and its principle depends on mathematical processing of time course of heat flow passing through the tested fabric due to different temperatures of bottom measuring plate (22 °C) and measuring head (32 °C) [17].

Thermal conductivity is a measure of the rate at which heat is transferred through unit area of the fabric across unit thickness under a specified temperature gradient [15, 30]. It is a property of a material [52], which depends on many factors such as fabric density, fabric porosity and fibre arrangement.

Regarding to the test results, the evaluated fabrics had higher thermal conductivity values after desizing (A-D) process. The statistical analysis showed that the desizing process had significant influence on the thermal conductivity for Feretiko ($p = 0,000$) and for Ayancik linen ($p = 0,001$) whereas the difference was not statistically significant for Burumcuk ($p = 0,105$). The wrinkle effect of Burumcuk fabric reduced after desizing process and it might be the reason of having less difference than Feretiko and Ayancik linen.

3.5 Thermal Resistance

Thermal resistance is an important parameter to understand the thermal insulation of a clothing. It is the ratio of the temperature difference between the two faces of a material to the rate of heat flow per unit area [30, 37], namely, refers to the ability of a system to resist heat flow [52]. Thus, the higher the thermal resistance, the lower the heat flow. Fabric thickness is one of the main factors influencing the thermal behaviour of the fabric. The increase in the thickness of a fabric increases its thermal insulation value due to the corresponding increase of the fabric volume and the surface area [46], which resulted by an increase in the amount of air trapped in the gaps between fibres and the yarns.

Table 4 indicates that for both before and after desizing, the thermal resistance of Feretiko was lower than other fabrics due to its lowest fabric thickness. Before desizing, Burumcuk had the greatest thermal resistance value whereas after desizing Ayancik linen presented higher thermal resistance results than Burumcuk. This might be caused by the change in the amount of air in the fabric interstices after desizing process. The wrinkle effect of Burumcuk leads to larger gaps on the fabric surface area so that the air movement could take place within them. After desizing process, Burumcuk had less wrinkle effect and therefore had lower fabric thickness, whereas Ayancik linen had the greater fabric thickness and so the greater thermal resistance values. Figure 10 shows the thermal resistance values versus the fabric thickness values regarding the fabric weights of Feretiko, Ayancik linen and Burumcuk fabric, respectively. It has been once again proved that the thermal resistance is proportional to the fabric thickness. According to the statistical analysis, in terms of desizing process, the thermal resistance values were statistically significant for Ayancik linen ($p = 0,000$) and for Burumcuk ($p = 0,035$) whereas were not statistically significant for Feretiko ($p = 0,672$).

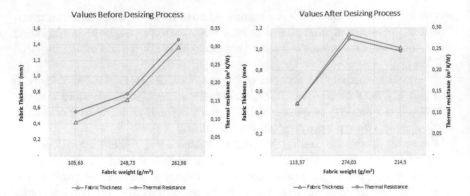

Fig. 10 The thermal resistance versus the fabric thickness for before desizing process (left) and after desizing process (right)

3.6 Thermal Absorptivity

Thermal absorptivity characterizes the transient thermal feeling during the first contact of a fabric with the human skin [15, 30], namely, determines the warm-cool feeling. It is the quantity of heat penetrating to a fabric from the skin during the immediate contact of the fabric with the skin. The higher the thermal absorptivity of the fabric, the cooler the feeling. The surface property of a fabric has great importance on thermal absorptivity, more uniform, flat and smooth fabric surface increases the contact area with the skin and gives cooler feeling [17]. On the contrary, rougher surface with lower regularity gives warmer feeling with a lower thermal absorptivity value.

With respect to Table 4, Ayancik linen had higher thermal absorptivity, in other words, had higher contact area with the measuring head of Alambeta, which presented the coolest feeling between evaluated fabrics before desizing process. After desizing, Feretiko showed the highest cooling feeling, due to the increase in the warp and weft density values (Fig. 11). Burumcuk had the lowest thermal absorptivity values for

Fig. 11 The thermal absorptivity values of evaluated fabrics

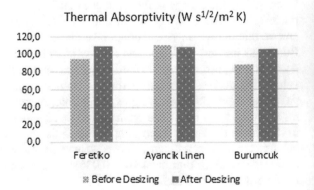

Thermal Absorptivity (W $s^{1/2}$/m² K)

Fig. 12 The air permeability
values of evaluated fabrics

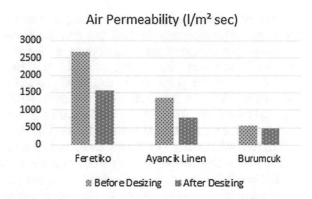

Air Permeability (l/m² sec)

both before and after desizing; the rougher surface due to the wrinkle effect decreases
the contact area and therefore the thermal absorptivity.

3.7 Air Permeability

Investigating air permeability is of importance for thermal comfort evaluation. It
is the ability of air to flow through the fabric [6] and it is defined as the measure
of airflow passing through a given area of fabric, at a given pressure and during a
time period [15, 30]. As many researches in the literature revealed [6, 10, 29, 31],
the fabric structure has a significant effect on air permeability, especially the pore
characteristics of fabric. Higher porosity therefore means more openness which gives
higher air permeability results.

The air permeability tests were performed using a Textest FX 3300 air permeability
instrument (Textest Instruments AG, Switzerland). Table 4 shows Feretiko had the
greatest air permeability results for both states, before and after desizing. Feretiko had
the lowest fabric thickness, fabric weight as well as the yarn number and this might be
attributed to the high porosity. After desizing process, the warp and weft densities of
evaluated fabrics increased, and as a result the porosity of fabrics decreased (Fig. 12).
Therefore, the fabrics had lower air permeability results after desizing process and
there was a significant difference on air permeability values of Feretiko (p = 0,000),
Ayancik linen (p = 0,000) and Burumcuk (p = 0,005) regarding desizing.

3.8 Clothing Insulation

In traditional methods, the thermal insulation of a fabric is measured with a heated hot
plate; however, the thermal insulation of a fabric is of limited value for evaluating

the thermal property when it is constructed into an ensemble. Therefore, thermal manikins, which measures the heat flux over the whole body surface area [18], is inevitable to be used to evaluate the clothing insulation. Thus, the thermal properties of these fabrics were tested using a thermal manikin, which enables to evaluate the thermal comfort properties in simulated conditions of use. For this purpose, long sleeve shirts with front buttons were manufactured. The garment patterns were prepared regarding to the body measurements of the thermal manikin and the shirts were produced using the same patterns for each fabric (Fig. 13).

The thermal manikin (PT-Teknik made in Denmark) used in this research is a female model that represents the similar dimensions with an adult woman. The thermal manikin only measures the dry heat transfer and in total has 20 thermally independent body segments (Fig. 14). The tests were conducted placing the manikin inside a climatic chamber, to provide steady environmental conditions during the tests. The testing conditions were specified according to the annual average temperature and relative humidity data of Black Sea Region, the hottest month was chosen and the conditions were determined as 23 °C ± 1 and 75 ± 5% RH.

Fig. 13 The thermal manikin with cotton shirt after desizing process

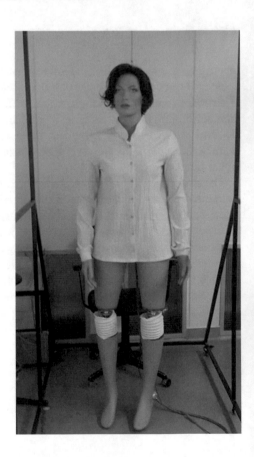

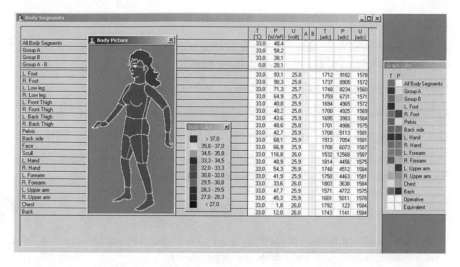

Fig. 14 The body segments of the thermal manikin

The manikin was placed around 0.1 m above the floor with hanging arms and legs and all the tests were set and conducted according to ISO 15831 standard. The trials were performed in constant skin temperature mode and the skin temperature of thermal manikin was set at 33 ± 0, 2 °C regarding to ISO 9920. The testing time of each measurement was 30 min, in which the skin temperature and heat loss were monitored every minute. At the end of each measurement, the heat loss was obtained and recorded to the computer.

The clothing ensemble's thermal insulation (I_{cle}) can be calculated by relating the heat flow from the manikin's surface area through the clothing into the ambient air with the heat flow from the nude manikin's body surface area, considering the temperature difference between the manikin's skin surface and the ambient air (ISO15831). Therefore, in order to determine the thermal insulation of the boundary air layer (I_a), tests with nude manikin were conducted under same static conditions. There are several methods to calculate the thermal clothing insulation (I_T). [32] compared global, serial and parallel methods. They found out that the serial method gives the highest values, whereas the parallel method always presents the lower values. Thus, in the present research, the global method of calculation of the thermal insulation was chosen.

In global method, the heat losses and the skin temperatures are calculated for each body segment with respect to the surface area factor (f_i) and afterwards are summed up. Therefore, the heat loss and skin temperature values are calculated for whole body before the thermal insulation is calculated. The equation for global method was presented in Eq. 2 [32], ISO 15831; [47]:

$$I_T = \frac{\sum_i (f_{i^x} t_{si}) - t_0}{\sum (f_{i^x} Q_{si})} \qquad (2)$$

where:

f_i—relationship between the surface area of the segment "i" of the manikin (A_i) and the total surface area of the manikin A ($f_i = A_i/A$).

t_0—the air temperature within the climatic chamber [°C].

t_{si}—skin surface temperature of the body segment "i" of the manikin [°C].

Q_{si}—sensible heat flux of the manikin obtained by area (W/m^2).

In order to calculate the effective clothing insulation (I_{cle}), Eq. 3 is used. As it has been mentioned above, I_{cle} is the difference between I_T and I_a:

$$I_{cle} = I_T - I_a \tag{3}$$

The test results calculated according to the global method were presented in Table 5 for before desizing (B-D) and after desizing (A-D).

The effective clothing insulation values of all fabrics behaved exact opposite to the thermal resistance values (Fig. 15). After desizing process, the effective clothing

Table 5 The thermal insulation and the effective clothing insulation values of evaluated fabrics

	I_T (m^{2}K/W)		I_a (m^{2}K/W)		I_{cle} (m^{2}K/W)	
	B-D	A-D	B-D	A-D	B-D	A-D
Feretiko	0,133	0,132	0,0947	0,0987	0,0388	0,0334
Ayancik linen	0,142	0,1416	0,0947	0,0987	0,0474	0,0429
Burumcuk	0,135	0,1392	0,0947	0,0987	0,0404	0,0405

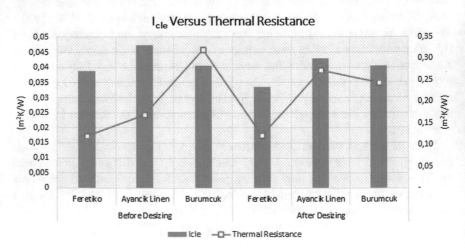

Fig. 15 Effective clothing insulation (I_{cle}) versus thermal resistance values

insulation values of Feretiko and Ayancik linen were lower whereas the result of Burumcuk was higher. The statistical analysis results showed that there were significant differences between the effective clothing results regarding desizing process (p < 0.05).

Burumcuk showed reverse I_{cle} results after desizing process and this might be caused by the decrease in its wrinkle effect. In order to calculate the effective clothing insulation, the air layer insulation (I_a) is subtracted from I_T, which is determined by nude manikin tests. By the change in wrinkle effect of Burumcuk, less air layer trapped between the garment and the body, and therefore the insulation generated more with the fabric than the air layer.

4 Conclusion

Throughout history, people in Anatolia had used handlooms to produce fabrics. There were many traditional fabrics in Anatolia where the raw material, weaving type and the applied processes vary regionally. These fabrics were produced and used since the Ottoman Empire; besides the population, the Ottoman sultans used these fabrics due to their thermal behaviours. Within the advances in technology, some of these fabrics are lost today; however, thanks to the local municipalities and community centres, some of them are still produced by traditional methods.

The traditional fabrics investigated in this chapter are Feretiko (Rize fabric), Ayancik linen and Burumcuk fabric. Feretiko is a local fabric of Rize City, which is located in the region of the Black Sea [12], is woven from hemp fibres and is produced at handlooms with whips, which is a setup made by ropes connected with the shuttle [9]. Ayancık linen is as well a local fabric produced as linen weft and cotton warp on traditional handlooms using the linen grown in the Middle Black Sea Region, which has a geographical indication [14]. On the other hand, Burumcuk is a fabric produced under different names in many regions of Anatolia and with different raw materials, namely, cotton, silk, linen or their mixtures. Although it was mostly produced using silk in the past, cotton became more widely over time. Burumcuk fabric is woven from double spun cotton yarn on narrow handlooms and afterwards the surface is folded by boiling technique to give the wrinkle effect [23, 38]. The Burumcuk fabric analysed in this study was 100% cotton produced in traditional handlooms in Sinop City.

The thermal properties of Feretiko, Ayancik linen and Burumcuk were tested before and after enzymatic desizing process in order to determine the effect of starch desizing on thermal properties. Thermal resistance, thermal conductivity, thermal absorptivity, air permeability, wetting and effective clothing insulation properties were investigated.

The wettability is about wetting the fibre surface by a liquid and knowing about the material's wettability especially in hot environments which is an outstanding factor to evaluate the clothing comfort. As it was expected, the contact angle values of all fabrics were lower after desizing process.

Fabric thickness is one of the main factors influencing the thermal behaviour of the fabric, and therefore the highest thermal resistance value is obtained from Burumcuk fabric before desizing process. After desizing, Ayancik linen had the greatest thermal resistance value as well as fabric thickness value.

The fabric structure is an important factor to evaluate air permeability, especially the pore characteristics of fabric. It was observed that Feretiko had the greatest air permeability results for both states, before and after desizing, due to its high porosity with the lowest fabric thickness, fabric weight as well as the yarn number.

Thermal comfort properties of fabrics during actual use may differ due to the conversion of two-dimensional fabrics to three-dimensional form. Especially, the air layer between the garment and the body surface provides additional insulation; also, the fit issues change the dynamics of the fabric surface. Therefore, the manufactured shirts for all fabrics were tested with a female thermal manikin. Regarding the effective clothing insulation, Feretiko, Ayancik linen and Burumcuk behaved exact opposite to the thermal resistance values. After desizing process, the effective clothing insulation values of Feretiko and Ayancik linen were lower whereas the result of Burumcuk was higher. The wrinkle effect of Burumcuk lead to larger gaps on the fabric surface area so that the air movement could take place within them. After desizing process, Burumcuk had less wrinkle effect and therefore had lower gaps on its surface.

Since Feretiko, Ayancik linen and Burumcuk are manufactured traditionally in handlooms, the yarns and the constructions are not as uniform as the fabrics manufactured industrially. Therefore, the results presented in this book chapter are to understand the thermal behaviours of traditional fabrics in order to contribute to the cultural continuity and sustainability.

Acknowledgements The authors gratefully acknowledge the funding by the project UID/CTM/00264/2019 of 2C2T – Centro de Ciência e Tecnologia Têxtil, funded by National Founds through FCT/MCTES. Derya Tama thanks FCT for fellowship 2C2T-BPD-08-2017.

References

1. Akın Z (2005) Keten Üretimini ve Keten Ürünlerini Tanıtım Bülteni/ Linen Production and Linen Products Promotion Bulletin. Ayancık Halk Eğitim Merkezi ve A.S.O. Müdürlüğü, Ayancık, Sinop. https://ayancikhem.meb.k12.tr/meb_iys_dosyalar/57/02/197883/dosyalar/2015_02/24045446_ayanckta_keten.pdf?CHK=ab957b6071897568ce5e2cea603cf838
2. Atalayer G (2012) Anadolu'da Yaşayan Dokumacılığın İzleri Üzerine/ On the Traces of Weaving Living in Anatolia. Paper Presented at The International Regional Textiles Congress, İstanbul, 2012
3. Atlıhan Ş (2016) Kuzeybatı Anadolu'da Bezlerin Kullanım Alanları ve Desenleri/ Uses and Patterns of Cloth in Northwestern Anatolia. In: Akalın M et al (eds) Proceedings of 7th international Istanbul textile conference on fabrics touching Anatolia, Istanbul
4. Başaran FN (2018) Anadolu Geleneksel Bez Dokumacılığından Bazı Örnekler ve Günümüzdeki Durumu/ Some examples of Anatolian traditional hand weaving and present situation. Arış Dergisi 13:14–25

5. Berber ŞG (2016) Konya İli Hadim İlçesi Hadim Dokumaları ve Son Dokuma Ustası Müjgan Akdemir/ Konya Province Hadim District Hadim Weaving and Last Weaving wright Müjgan Akdemir. In: Akalın M et al (eds) Proceedings of 7th international Istanbul textile conference on fabrics touching Anatolia, Istanbul

6. Bivainyte A, Mikucioniene D (2011) Investigation on the air and water vapour permeability of double-layered weft knitted fabrics. Fibres Text Eastern Eur 19(3):69–73

7. Büken ORN (2005) El Dokumacılığının ve El Dokuma Tezgahının Tarihçesi, El Dokuma Tezgahı Çeşitleri/The history of hand-weaving, hand-weave loom, Kinds of Hand-Weave Loom. Sanat Dergisi 5:63–84

8. Caschera D, Mezzi A, Cerri L et al (2014) Effects of plasma treatments for improving extreme wettability behaviour of cotton fabrics. Cellulose 21:741–756

9. Cavus A (2016) The production of Rize fabric (Feretiko) as a traditional handicraft and touristic product. Int J Acad Res Bus Soc Sci 6(4):189–200. https://doi.org/10.6007/IJARBSS/v6-i4/2089

10. Ciukas R, Abramaviciute J (2010) Investigation of the air permeability of socks knitted from yarns with peculiar properties. Fibres Text Eastern Eur 18(1):84–88

11. Dölen E (1992) Tekstil Tarihi - Dünyada ve Türkiye'de Tekstil Teknolojisinin ve Sanayinin Tarihsel Gelişimi, Marmara Ünv. Teknik Eğt. Yay: 92/1, İstanbul

12. Er B (2018) Feretiko (Rize clothing) as a cultural identity of Rize city. J Int Soc Res. 11(59):517–527. http://dx.doi.org/https://doi.org/10.17719/jisr.2018.2657

13. Evecen A, Beydiz MG (2018) Paleolitik ve Neolitik Dönem Buluntularında Giyim Kültürü/ Garment culture in the foundlings of paleolithic and neolithic age. Turkish Studies Social Sciences Volume 13/10, Spring 2018:303–333

14. G.I. (2020) https://www.ci.gov.tr/Files/GeographicalSigns/8d89b568-83c3-4502-9840-8d8def cf9c4d.pdf. Accessed 22 Sept 2020

15. Gidik H, Bedek G, Dupont D (2016) Developing thermophysical sensors with textile auxiliary wall. In: Koncar V (ed) Smart Textiles and Their Applications, Woodhead Publishing Series in Textiles, p 722

16. Güngör F (2016) Maut Yöntemi İle 'Ayancık Keteni'nin Şap Mordanına Göre Uygun Boyama Reçetesinin Belirlenmesi/ Determination of the appropriate dyeing prescription of "Ayancık Linen" according to alum mordant with Maut method. In: Akalın M et al. (eds) In: Proceedings of 7th International Istanbul textile conference on fabrics touching Anatolia, Istanbul

17. Hes L, Loghin C (2009) Heat, moisture and air transfer properties of selected woven fabric in wet state. J Fiber Bioeng Inform 2:141–149

18. Holmear I (2005) Protection against cold. In: Shishoo R (ed) Textiles in sport. Woodhead Publishing Limited, pp 262–286

19. Hu JL, Lu J (2016) Memory Polymer Coatings for Smart Textiles. In: Hu J (ed) Active coatings for smart textiles, Woodhead publishing series in textiles, USA

20. ISO 15831:2004, Clothing-Physiological effects-Measurement of thermal insulation by means of a thermal manikin.

21. ISO 9920:2007: Ergonomics of the thermal environment-Estimation of thermal insulation and water vapour resistance of a clothing ensemble

22. Karaoğlan H (2017) M.Ö 2000'de Anadolu'da Kumaş Üretimi (Arkeolojik Buluntular Işığında)/ Fabric Production In Anatolia BC 2000 (In the light of archaeological findings). J Soc Sci Inst-Afro-Eurasia Spec Issue 103–116

23. Koca E, Vural T (2013) The shirts used in Turkish folk dress. In: Paper presented at the 7th international Turkish culture, art and cultural heritage preservation symposium/art events, Baku-Azerbaijan, 26–29 June 2013

24. Komşuoğlu Ş, İmer A, Seçki M (1986) Resim II/ Moda Resmi ve Giyim Tarihi. MilliEğitimBakanlığı, Ankara

25. Korkmaz E (2016) Denizli İli Yaşayan El Sanatları/ Living Crafts of Denizli Province. Master's Thesis, Süleyman Demirel University, Isparta

26. MEGEP (2012) http://megep.meb.gov.tr/mte_program_modul/moduller_pdf/%C5%9Eile% 20Bezi%20Dokumaya%20Haz%C4%B1rl%C4%B1k.pdf. Accessed January 2021

27. Mehta R (2018) Experimental study on application of different sizing agents and its impact of fabric properties. Int J Multidiscip Educ Res 1(4):188–198
28. Merev TA (2019) Tarihi Tekstillerde Bozulma Nedenleri ve Restorasyon Öncesi Yapılması Gereken İşlemler / Causes of deterioration in historical textiles, and procedures to be prepared before restoration, Fatih Sultan Mehmet University. Master's Thesis, İstanbul
29. Mezarcıöz S, Mezarcıöz S, Oğulata R (2014) Prediction of air permeability of knitted fabrics by means of computational fluid dynamics. Textile Apparel 24(2):202–211
30. Mishra R, Militky J, Venkataraman M (2019) Nanoporous materials. In: Mishra R, Militky J (eds) Nanotechnology in textiles; theory and application. Woodhead Publishing, p 399
31. Ogulata RT, Mavruz S (2010) Investigation of porosity and air permeability values of plain knitted fabrics. Fibres Text Eastern Eur 18(5):71–75
32. Oliveira AVM, Gaspar AR, Quintela DA (2008) Measurements of clothing insulation with a thermal manikin operating under the thermal comfort regulation mode: comparative analysis of the calculation methods. Eur J Appl Physiol 104:679–688
33. Oksenvag JHC, Fossen M, Farooq U (2019) Study on how oil type and weathering of crude oils affect interaction with sea ice and polyethylene skimmer material. Mar Pollut Bull 145:306–315
34. Özomay M (2016) Türkiye'de Yöresel Dokunan Bez Örneklerinin Doğal Boyarmaddeler İle Griİlişkisel Analiz Yöntemi Kullanılarak Boyama Özelliklerinin Belirlenmesi/ Determination of dyeing properties of cloth samples locally woven in Turkey using grey relational analysis method, Phd Thesis, Marmara University, İstanbul
35. Öztürk İ (2016) Yöresel Bez Dokumaların Değişime Uğraması, İşlev Değiştirmesi, Kaybolması/ The Change, Change of Function, and Disappearance of Local Cloth Weaving. In: Akalın M et al (eds) In: Proceedings of 7th international Istanbul textile conference on fabrics touching Anatolia, Istanbul
36. Saatçioğlu K (2012) Geleneksel Türk El Dokumalarından Rize Bezi'nin (Feretiko) Süsleme Teknikleri İle Giysiye Uyarlanması/ The ornamentation technics with rize cloth (Feretiko), One Of The Traditional Turkish Hand Woven Fabrics, Together With Its Adaptation to Garments. Master's Thesis, Haliç University, İstanbul
37. Salman F (2010) Türk Kumaş Sanatında Görülen Geleneksel Kumaş Çeşitlerimiz/ The Traditional Names In The Art of Traditional Turkish Fabric. Sanat Dergisi 0(6):13–42
38. Salman F (2011) Türk Kumaş Sanatı, Zafer Ofset Matbaacılık, Erzurum
39. Saville BP (1999) Physical testing of textiles, Woodhead Publishing Series in Textiles, p 314
40. Schellbach SL, Monteiro SN, Drelich JW (2016) A novel method for contact angle measurements on natural fibers. Mater Lett 164:599–604
41. Seçkinöz M, Alpaslan S, Komşuoğlu Ş et al (1986) Resim II/ Süsleme Resmi ve Süsleme Sanatları Tarihi. Devlet Kitapları, Türk Tarih Kurumu Basımevi, Ankara
42. Selçuk K, Yurttaş H (2019) Rize Feretiko (Rize Bezi) ve Bayburt Ehram Dokumacılığı Üzerine Bir Deneme / An Experiment on Bayburt Ehram and Rize Feretico (Rize Cloth) Weaving. J Turk Res Inst TAED-66 539–559
43. Selçuk K, Yurttaş H (2020) Doğu Karadeniz Bölgesi'nde Dokunan Geleneksel Kumaşlar Üzerine Bir Deneme/ A trial on traditional fabrics woven in the eastern black sea region. Karadeniz Araştırmaları Enstitüsü Dergisi 6(9):35–47
44. Sipahioğlu O (1992) Bursa ve İstanbul'da Dokunan ve Giyimde Kullanılan 17. Yüzyıl Saray Kumaşlarının Yozlaşma Nedenleri/ The causes of corruption of 17th century palace fabrics woven and used in clothing in bursa and Istanbul. Master Thesis, Dokuz Eylül University, İzmir
45. Sözen M, Güner Ş (1998) Türk El Sanatları. Hürriyet Gazetecilik ve Matbaacılık, İstanbul
46. Stankovic SB, Popovic D, Poparic GB (2008) Thermal properties of textile fabrics made of natural and regenerated cellulose fibers. Polym Testing 27(1):41–48
47. Tama D, Catarino A, Abreu MJ (2019) Evaluating the effect of water-repellent finishing on thermal insulation properties of rowing shirts using a thermal manikin. Text Apparel 29(4):279–288
48. Tüm Cebeci D (2019) Peşkir ve Mahrama Dokumalarının Sinop El Dokumacılığındaki Yeri/ The Place of Peşkir and Mahrama Weavings in Sinop Hand Weaving. İdil, 61 (2019 Eylül):1187–1198

49. Türktaş Z (2016) Konya İli Hadim İlçesi'nde Bez Dokumacılığı ve Günümüzde Devam Eden Bir Faaliyet Örneği; Müjgan Akdemir/ Cloth Weaving In Konya Province Hadim District And An Example Of Ongoing Activity Today; Müjgan Akdemir, In: Akalın M et al (eds) In: Proceedings of 7th international Istanbul textile conference on fabrics touching Anatolia, Istanbul
50. Tütüncüler Ö (2006) Çorum-Resuloğlu Eski Tunç Çağı Mezarlığı'nda Kumaş Kullanımına İlişkin Yeni Bulgular/ New findings regarding the use of fabrics in Çorum-Resuloğlu old bronze age cemetery. Anatolia 30:137–148
51. Uysal ÖB (2016) Anadolu'da Gözenekli Yapı İle Elde Edilen Desenli Kumaşlar/ The Patterned Fabric Obtained By The Porous Structure In Anatolia. In: Akalın M et al (eds) In: Proceedings of 7th international Istanbul textile conference on fabrics touching Anatolia, Istanbul
52. Uzumcu MB, Sari B, Oglakcioglu N et al (2019) Comfort properties of silk/cotton blended fabrics. Fibers Polym 20(11):2342–2347
53. Yağan ŞY (1978) Türk El Dokumacılığı. Türkiye İş Bankası Kültür Yayınları, İstanbul
54. Yuan Y, Lee TR (2013) Contact angle and wetting properties. In: Bracco G, Holst B (eds) Surface science techniques, springer series in surface sciences 51. Springer-Verlag, Berlin, Heidelberg, pp 3–34

Indian Handloom Design Innovations and Interventions Through Sustenance Lens

V. Nithyaprakash, S. Niveathitha, and V. Shanmugapriya

Abstract In this chapter the status of Indian handlooms in the current scenario is discussed with reference to the design attributes of the Kanjivaram, Ikat, and Jamdani sarees, followed by the discussion on the changing face of Indian fashion semiology. The design innovations and interventions across these three categories are analyzed from the perspective of the cultural sustainability of the handloom design palette and socio-economic conditions. Along the course of discussing the design innovations and interventions, the direct and indirect implications of the technology or market driven modifications are explained. Further the scope of loom design techniques and pre-loom design techniques, design approach, and product design diversification has been elicited. This investigation justifies the role of design interventions with reference to the prospects of augmenting weaving craftsmen's artisan skills and improvising the handloom design palette according to the new age consumer needs. Subsequently the impact of design intervention strategies on weaving craftsmen socio-economic conditions are categorically analyzed. This case study analysis concludes the effective design interventions in the manufacturing practice of the chosen Indian handloom products.

Keywords Design intervention · Design palette · Weaving craftsmen · Artisan · Motifs · Ornamental rendering

V. Nithyaprakash (✉)
Department of Fashion Technology, Kumaraguru College of Technology, Coimbatore, Tamil Nadu, India
e-mail: nithyaprakash.v.ft@kct.ac.in

S. Niveathitha · V. Shanmugapriya
Department of Fashion Technology, Bannari Amman Institute of Technology, Sathyamangalam, Tamil Nadu, India
e-mail: Niveathitha@bitsathy.ac.in

V. Shanmugapriya
e-mail: shanmugapriya@bitsathy.ac.in

1 Introduction

Indian Handloom products account for 95% of the World's Global Handloom product [1], and they reflect the woven treasure house of Indian textiles. Clothing articles among the handloom designs include Sarees, Mekhala Chaddar, Shawls, Stole, Mufflers, scarf, Angavastram, Dhoti, Lungi, and Sarongs [2]. Among the handloom made clothing products, sarees account for 89.4% of the total goods produced out of it [3]. The handloom sarees are manufactured in both frame looms and pit looms. Frame looms account for 31.5% and pit looms enumerate up to 42.2% of the total looms [2]. Out of the total number of pitlooms, close to 67% of them are fitted with either dobby or jacquard mechanism, which implies high end clothing articles are prepared out of them [2]. However, the design palette extensively varies from one region to another region. And every weaving cluster drawing from its own history and legendry, specializes in producing a unique visual imagery and design palette. Further the handloom saree products can be broadly classified into two categories of products with reference to the level of design complexity comprised in it. Products with an all over design layout comprising intricate motifs, ornamental figures, exquisite pallu, and ornamental borders are indicated as complex designs. The handloom products in this category are largely confined to expensive sarees tagged under the design palette of Banarasi, Jamdani, Kanjivaram, Paithani, and Patola.

The Handloom saree designs are imparted in two different ways; Pre-loom design work and On-loom design work. Handloom designs produced by Ikat patterns conform to pre-loom design technique and those designs applied on-loom are produced through extra warp or extra weft interweaving techniques. The elements such as color, thickness and appearance of yarn, surface texture along with ornamental rendering style ascribes the socio-cultural values of handloom designs and the repertoire of the design palette. These high value complex design sarees are produced from pure silk or high-end cotton and gold zari work. Pure zari refers to composition of 98.5% silver coated with 24 carat Gold [4]. The semiotic value of these complex design palette connotes Aristocratic and Elite values in Indian fashion which requires 4 weeks to even 36 months' time frame weighing in by the intricacy of the design palette engaged by the weaving craftsmen. Generally, the vintage weave and craft is preserved while the color and motif style undergo fashion adoption strategies.

The second category designs include sarees worked and produced with dobby/jacquard and extra weft techniques, but the design palette is more toned down in terms of the design layout comprising fewer motifs and minimalistic ornamental work ascribed to the budget prices. These designs are made from silk blends and textured zari; a cheaper variant of zari where the copper threads are covered with gold layer instead of the expensive silver. Both complex designs and moderate designs uses finer yarn above 80 s count while the Kanjivaram varieties use 2 ply yarns which makes the fabric heavy. According to the Fourth Handloom Census, Cotton is the most sought-after yarn type among 65.2% of the weavers. The silk, cotton and its blends are preferred by about 7% of the workers especially for moderate to high value articles [2].

1.1   Challenges of Indian Handloom Saree Designs

The challenges faced by Indian handloom sarees are lack of contemporary designs in handloom products [5], unwillingness of the younger generation to take up ancestral crafts [6], Speculative raw material prices, and lack of fair-trade culture. Among the pioneering efforts for the reigniting the passion for woven crafts among the children and the second-generation weavers, Shailin smith, Director of The Raj group imparts education to the children and manual training [6]. Moreover, the onslaught of pandemic COVID-19 has resulted in reverse migration of labor of the weaver folk back to their rural households. And with the Indian economy realigning its strategies to manufacture locally and market in the domestic sector in the name of "Atmanirbhar Bharat", there has been policy changes to regulate the raw material prices of handloom products [7]. As an after shoot of new government strategies, services related to training, distribution of looms and accessories along with providing yarn supplies has been extended to the handloom weaving clusters with renewed vigor [7]. Though as early as 1982, Dastkari bazaars were found to address this gap between urban consumers and weaving craftsmen in terms of providing design assistance and market requirements [8], still there is widespread lacuna in deciphering and translating the contemporary design sensibilities to the craftsmen according to the latest working paper published by Export and Import bank of India in 2018 [2].

In this study, the changing face of Indian fashion semiology, design interventions in handloom sarees and the influence of technological innovations in handloom process with emphasis on improving weaving craftsmen or artisans' livelihoods through income generation and preservation of cultural values of handloom products has been examined. The methodology followed is a combination of primary research involving discussions with scientists of government agencies such as Central silk board that explicates the weaving techniques and secondary research, involving study about the traditional design palette of Kanjivaram, Ikat, and Jamdani sarees, study of Indian fashion semiology, case study discussions of design intervention strategies deployed across these aforementioned sarees vis a vis the technological innovations in their process. The kinds of ideas and lifestyle values that this changing face of Indian fashion semiology aids to infuse design intervention strategies are also elaborated.

2   Design and Making of Traditional Kanjivaram, Ikat, and Jamdani Sarees

2.1   Kanjivaram Sarees

The Kanjivaram motifs draw inspiration from the Architecture and paintings of the Temples gracing the Kanchipuram city comprising of peacock, swan, rudraksh seed, yazhi (mythological creature), parrots, jasmine buds, gandaberuda (two headed bird), creeper vines, paisley or mango, rings, stripes, musical anklets used for dance, Elephant, Lotus, lion, vase of plenty (Poorna Kumbha), and tree of life (kalpa Vruskha) [9]. What makes Kanjivaram special is its rich, ornamental rendering of these motifs in chosen color backgrounds with intricately made border of contrast color that sports off an ornamental grandeur look. The checks and stripes of Kanjivaram saree runs into an extensive palette analogous to the tartan plaids and checks of Scots. A group of checks seen in the Kanjivarams include many grid types of different modular spaces known by different names such as Paalum and Pazhamum alias 'milk and fruit' engaging auspicious colors of red, yellow, and green [9]. The family of stripes present in Kanjivaram range from narrow blocks of stripes in moderate contrast colors referred as 'plantain leaf stripes', a broader block of stripes demarcated by thin double zari lines; aka railway cross stripes, wavy stripes with dots between them; aka Neli stripes to fine needle like lines; aka 'vaira oosi' [9]. Meanwhile the family of check includes intricate diamond checks to simple square check boxes of desired modular spaces. Apart from these stripes and checks, it has intricated complex geometric patterns formed by lines and shapes mimicking the traditional bridal hairdo of Tamil Nadu. The distinctive pattern of arrangement makes it look like a grid work inscribed on Gold jewels.

The color palette of Kanjivaram saree draws from the cultural roots of mythology and folklore icons and symbols with clear boundaries ranging from jewel tones to smooth hues [10]. The Kanjivaram color palette is home to sixteen hues of green, sixteen tones and shades of blue, seven hues of pink, five shades of red, five shades of yellow, three hues of orange, and the gold tone. Thus, it is deemed diverse among the styles of weaving owing to the unique versatile vocabulary of the design elements [10].

2.1.1 The Anatomy of Kanjivaram On-Loom Warp Color Arrangement

Petni, kondi, and Reku represent the traditional handloom weaving techniques for preparing pallu, border design, and body colors across the length of the fabric, sarees [11]. They arrange the warp threads in necessary color pattern across defined width. In the Kanjivaram sarees, the border and pallu are of same color and brightly contrast the body color. Solid color is obtained by weaving warp and weft in same color. The first half meter to three fourth meter length out of 5.5 m is described as pallu. Usually, the pallu is solid colored over which extra warp or extra weft designs are produced. The border is marked around the edges of selvedge about 3″ to 9″ width.

The warp threads are dyed in two colors, the starting length marked around half a meter to three forth meter is allocated for pallu color. And the remaining warp length is dyed in contrast color. So, the same set of warp ends are dyed in dual color which are said to occupy the body section. Similarly, there are three sets of wefts: one for body, one for border, and one for pallu. To have two different colored warp, traditional weavers use three different techniques identified by the place of their origin. Petni is the name of the technique used in Kanchipuram for producing two colored warps.

For Petni work, three ball warp beams representing two borders on either side with body warp in center are gaited and set on the weaving machine. The body weave is continued as per the requirements for the body length as the weaving is carried out with three shuttles, two for weaving the border color, and one for weaving the body color. Meanwhile jacquard shedding or Jala technique along with heald frame shedding is also accommodated to incorporate extra warp designs. Once the body part weaving is completed, petni work starts. Now the pallu section colored warp end wound on separate ball warp is drawn, loop twisted and joined to body warp ends at the back of the healds. As the heald and reed are kept near the cloth fell, this petni work is comparable to the process of tying new warp ends to old warp ends. Now the heald and reed are slowly pulled toward the back to allow the joined ends reach the cloth fell. Now a lease rod is inserted in the space between cloth fell and the roller that allows the longer loops of joined ends to be untwisted and gaited properly. Now we have two ends in the space occupied by one warp end across the whole fabric width. Thus, weaving is continued slowly for around 2″ length. This portion of weaving is called petni weaving. Now again the reed and heald frames are moved upfront near cloth fell followed by rising the original body warp ends and lowering the pallu warp ends. Further the body warp ends are cut at the cloth fell and removed. Later the weaving of the pallu part is continued.

Ilkal of Karnataka uses a technique called kondi that traces its origin to eighth century AD to obtain two color warps: pallu and body [11]. Initially the body warp is prepared on a peg warp mechanism. Here the warp body length is 2.5 m and made from 2 ply warp yarns. The remaining warp length is covered by pallu warp covering the remaining 1 m. The total no of warp ends is double the number of ends representing the width of the fabric. The warp ends are drawn too and forth continuously forming an open loop structure. The structural arrangement has open loops at both sides. The continuous arrangement creates open loops at alternating ends in such a way the open loops formed at the even number ends are seen on one side and the open loops formed at odd number ends on the other side. The pallu warp wound on a bobbin is a single ply yarn of contrasting color is drawn through the open loop and dragged between the pegs. The pallu end dragged between the pegs is taken through the open loop of body warp from the right-side end and dragged further to the left peg and wrapped around it. This drawing operation creates the first interlocking between the open loop of body and pallu warp. The pallu warp end wrapped on the left peg is again extended up to right peg and kept ready for next interlocking task. Similarly, the pallu warp is drawn through all the open loops of warp back and forth using the peg mechanism that completes the interlocking of all body warp yarns and pallu warp yarns. This process is termed as kondi peg warping technique. After establishing the interlocking mechanism, the pallu warp loops are unmounted from the pegs and their parallel alignment of loops is maintained by placing a leasing thread in between them. One loop of body warp interlocked with one loop of pallu warp is assigned as one warp end. Each pallu warp loop interlocking the body warp loop comprises of two single ply threads which are twisted together to form 2 ply pallu warp end. Thus, the twisting operation of pallu warp ends permanently interlocks the body warp loop. The prepared kondi warp is now gaited and kept ready for weaving.

First the pallu part is woven with same color weft thread for around one-meter length. Along the pallu weaving process, required additional ornamentation via extra weft technique is also incorporated. When we come near interlocking area of body warp and pallu warp which covers around 1 ½ inches, weaving is slowed down as we have four ply body warp ends for every two ply pallu warp ends. Out of each four plies in warp end, two plies are cut at the fell of the cloth and further removed through the healds and reed thus extending the body warp length. The cut ends at the cloth fell are in the shape of U hooks. These "U shaped hooks" at the interlocking section of body warp and pallu warp earns the name "kondi" in Kannada. Kondi literally denotes bent shape in Kannada. And for the weft design work, tie and dye technique and korvai techniques are used for producing weft insertion colors.

2.2 Ikat

Patola renowned for its profusive intricate work in multiple color background undergoes a lengthy process starting with tying and dying of warp yarns and weft yarns then followed by weaving them. The choice of colors in the dyed regions of yarn requires careful planning to rule out its frequent appearance across the visual design palette. Unlike the representative depiction of motifs in Kanjivaram sarees, here the motifs are of abstract and geometric in appearance yet still inspired by the same repertoire of Elephants, parrots, Florals, Human figures, Shikars, and Paan [12]. The profusive intricate work takes a long month of dedicated craftsmen to complete it. Another distinguishing feature of Patola sarees visual design palette is its reversibility. The type of wrapping the yarns and dying them at desired places well before aligning them on the loom at designated locations is referred to as Ikat in Telangana and Andhra Pradesh [12]. When both warp and weft yarns undergo a dying process before weaving on looms, it is termed as Double Ikat and when only weft or warp yarns are dyed and woven into sarees they are referred as single Ikat. The more the dying involved in saree's visual design palette, the more the complex the task gets [12]. The design palette of the Patola comprises of as many as 11–12 hues [13]. Patola is capable of being made from jute, cotton, or synthetic blends apart from the traditional silk [13] (Fig. 1).

Patola saree designs are clustered under different design names implying the pattern type and color layout used in them. The Patola Ikat design comprises of several sub designs like Narikunj, Panpatola, Navaratna, Ambala, Teilya rumal, and Swastik each have distinctive visual design layout [14]. These are said to connote the socio-cultural values of Gujarat and patronized by Hindus, Jains, and Buddhists Alike [14]. Pochampally, a name in the tentative list of UNESCO heritage sites, is home to one of the oldest weaving clusters in India [15]. Comparable to Patola Ikat, Pochampally Ikat's visual design palette comprises of abstract patterns distributed all over the saree but the distinguishing feature between Patola Ikat and Pochampally Ikat caters to the grid layout with diamond shaped modular spaces where the geometrical elements are interspersed between them in the body part of the Pochampally design

Fig. 1 Patola Ikat saree design. *Source* (https://commons.wikimedia.org/wiki/FileDescription_Tex tile_artists_demonstrate_double_ikat_weaving_at_the_2002_Smithsonian_Folklife_Festival_feat uring_The_Silk_Road_(2,548,928,970).jpg)

palette [16]. The word 'Ikat' is derived from the Malayan word 'mengikat' [17] (Fig. 2).

Odisha's Ikat is even more intricate where even religious verses from the Jayadeva's Gita Govinda are rendered on the handloom woven fabrics [18]. Unlike the precise geometrical, rather stepped edge ornamental figure rendering style of Patola, the Odisha Ikat's ornamental figure rendering style deploys more curvilinear edges. The ornamental figures of Odisha Ikat encompasses all and sundry aspects of beauty defined in religious scriptures and literary works such as fluid movement posture of ducks, deer's attractive eyes, Elephant's smooth walking gait, Lion's slender waist, and Lotus flower's delicate silhouette [18]. The rendering style carries more poetical expression on fabric. Single Ikat with varying weft float length manipulations define the design palette of Odisha Ikat (Fig. 3).

2.2.1 Making of Ikat

Dyed warp yarn preparation in Ikat weaving undergoes a tedious process of 1. Transferring the yarn from cones to huge life size chakra and arranging them in several small groups of loose coil forms. These loose coil forms of warp yarns usually represent a unit design repeat of the final pattern for dyeing whose number in a group is again determined by the size of design unit. 2. The dyeing process begins with tying

Fig. 2 Pochampally Ikat Saree worn by classical dance exponent. *Source* (https://commons.wik imedia.org/wiki/FileKuchipudi_Dance_2.jpg)

the loose coil forms with rubber strips whose purpose is to resist the dye pickup, marking the places followed by dyeing with lighter colors first and continued until the darker colors get dyed. 3. After dyeing the warp yarns are aligned through the harness as per the lifting plan of the weave and the placement order of the unit design repeat [19]. The weft yarn dyeing of Ikat weaves is akin to the warp yarn dyeing except the transferring method. The weft yarns allocated for design pattern are transferred to pegs where they are grouped, marked with horizontal design unit of the pattern, tied, and dyed. Later the dyed weft yarns are wound on pirns that end up in the shuttle. Each weft yarn needs manual alignment according to the horizontal design unit of the pattern [19].

Fig. 3 Odisha Ikat saree design. *Source* (https://commons.wikimedia.org/wiki/File:Sambalpuri_S aree.jpg)

2.3 Jamdani and Uppada Jamdani

T.N. Mukarji refers Jamdani as Jamdani muslin comprising of geometric floral patterns and ornamental figures. Jamdani is a cotton woven fabric and represented one of the finest varieties of muslin [20]. The Jamdani patterns comprised of geometric, plant vines, and florals that are said to have originated through the fusion of Mughal and Persian art ensembles thousands of years ago and patronized by the Mughals and the aristocrats [20]. The elaborate motifs and patterns of Jamdani are directly woven into the fabric by extra weft interlace or interweave technique like the embroidery using small narrow shuttles housed with fine bobbins of gold, silvery, and multi-color threads [21]. In the earlier days, the master weavers rendered the patterns and ornamental directly on the loom reflecting from their memory. Later, the visual imagery of the patterns remained rendered on the graph sheet and placed under the loom while interweaving or interlacing them on the fabric ground (Fig. 4).

Uppada sarees, a well renowned name for their Jamdani work. Uppada sarees, once the staple clothing article of the Andhra Pradesh's Royal women is originally made from pure silk of light weight category with gold zari ornamentation of Jamdani Technique which gradually lost its fame [22]. However not until 1985, the art of Uppada Jamdani got its identity through Ghanshyam Sarode, who established the

Fig. 4 Jamdani saree design. *Source* (https://commons.wikimedia.org/wiki/File:Jamdani_DSC_0500.JPG)

design palette and the visual imagery according to the recitals of the Jamdani work and the design layout heard from the elite Mumbai customers [23].

Motifs and ornamental figures in Uppada Jamdani have an abstract outlook expressed as summative shapes of its real time inspiration. As the real time cultural symbols have organic shape, their reproduction process is cumbersome and slow. Hence the shape and silhouette of the real time cultural symbols are modified to suit the regular features of geometrical objects so that they are easily reproducible [24]. The outlines of the motifs appear to be wavy like an embroidered motif outline rendered using Holbein stitches. Real time inspiration for these motifs remains borrowed from animals, florals, and Mughlai art forms. Slowly motifs from the folklore were also added on [24]. One such style of Uppada Jamdani is butidar where the distribution of floral motifs can be seen all over the saree and rendered diagonally across the body of the saree.

The Shantipur Jamdani comprises of small checks, stripe effect rendered by colored threads with tie, and dye patterns on the pallu [25]. Other design variants of Jamdani include Dhaiakhali Jamdani that has a contrast border and Tangail Jamdani; the one with single tone border or dual tone border. Among all the Jamdani saree designs, the 'Panna Hazar' design is still the most coveted design for its connotation of socio-cultural values. The 'Panna Hazar' literally translates to thousand emeralds [25]. Nowadays Jamdani are made from machine made cotton yarns, as well as silk and silk-cotton blends.

3 The Changing Face of Indian Fashion Semiology

The design palette of the handloom fabrics always shared the socio-cultural properties expressed by the ornamental motifs and figures of humans, plants, and animals. The ornamental motifs and figures of humans, flora, and fauna served as basic signifiers among the handloom designs despite the very many rendering styles of each GI tagged weaving cluster. The rendering style of the ornamental figures found in handloom products draws an apparent resemblance to prehistoric Bhimbetka cave painting's diminutive figures that might have trans evolved and passed down the centuries through Art, Sculptures, Architecture, Poems, and Treatises. Archeologist Misra refers the diminutive painting figures to ritualistic and spiritual significance [26, 27]. The ornamental rendering style denoting the aesthetic virtues of beauty as defined in Indian classical literature are extolled to serve as symbols of religious and political hierarchies during medieval times [28]. According to the semiotics that prevailed during the medieval times, Ikat patterns on silk was considered luxury of the Gods and the Kings, while Ikat patterns on cotton was deemed to represent the common man and the working class [28]. The Indian clothing produced by handlooms acted upon as agents of holiness, purity, and spirit of gift rather than mere symbols of social and political status during the pre-colonial times [29].

Later in the 1970s saree draped in nivi style became the national dress, patronized by none other than the India's first woman prime minister Mrs. Indira Gandhi. The spurt in education, economic growth, financial independence, the rise of IT industry, entry of satellite channels in 1990s followed by access to global luxury products since 2000's through modern retail opened new avenues for accessing global fashion products to the Indian women [30]. The Indian woman evolved around the lifestyle changes in her own ways striking a balance between profession and family life. Suddenly the new extended roles of Indian women led them to embrace new western outfits: pantsuits, ties, and dresses over the traditional ethnic outfits [31]. Nevertheless, the emergence of Sushmita Sen, Aishwarya Rai, Priyanka Chopra, and Manushi Chillar as fashion icons has also brought in new normal for the definition of beauty which has led to more women adopting global fashion and luxury products. Further the new Indian women's financial independence and lifestyle has allowed her to explore saree kitsch fashion varieties such as sarees worn top of jeans or trousers and pleated knee length sarinis [32].

The popular western silhouettes for office wear of women constitute shirt waist dress, pant suits, skirt suits blouse, red carpet gown, trousers, cocktail dress, and evening gown. However, the emergence of new Ethnic Indian brands like Odhni, Koskii, etc. has brought in layered kurthas, lehengas, silk gowns, and Indo-western kurthis with repackaged presentation of traditional motifs and ornamental figures in contemporary styles [33]. These contemporized versions of ethnic silhouettes have given the postmodern Indian women options to exercise them for everyday office wear carrying forward the expression of ethnic roots [34]. But the Indian saree still holds a revered position and noted that designers like Sabyasachi Mukherjee, Gaurang Shah, Abraham and Thakore, Anavila Sindhu Mishra, etc. have given it a new interpretation to make it more relevant to the changing times and reinstate its status [32]. The major Indian festivals like Holi, Ganesh Utsav, Navratri /Durga Pujo, Diwali, etc. are occasions for family reunions and celebrations, where the women who have discontinued wearing sarees as a daily wear do don on gorgeous designer sarees [35]. Meanwhile, the likes of brand Tjori specializes in exporting Indian Artisanal and craft detailed multiple line products: Contemporized versions of Trench coats with Ikat patterns, Kashmiri shawls and stoles, Nehru jackets, Jacquard designed bags with embroidery, Capes, and wraps.

Developmental efforts by organizations such as Craft revival trust, NGO's, Handloom associations, Dastkari committees, etc. portray the fine craft, socio-cultural value chain of handloom products, and market them as Artistic goods of high value. What is imperative refers to sustaining and improvising these handloom designs and the people associated with them by accommodating the new design preferences led by lifestyle changes. In other words, adopting contemporary design sensibilities that are capable of harnessing socio-economic benefits for weavers simultaneously preserving the cultural heritage of artisan skills deems crucial for the hour. Contemporary design sensibilities encompass new visual communication attributes such as deconstruction of tradition, deconstruction of graphic, deconstruction of materials, and use of new color contrasts.

4   Design Interventions in Indian Handlooms

Designer products are works of creative aesthetic expression which are highly valued for their design palette and rendering style. The onus of improving the livelihoods of craftsmen through the entrusted task of creating successful designs solely lies with the designer. The word design applies to creation of not only material objects but also inclusive to the production of successful design values strategically through innovative products and services [36]. Thus, design is perceived as an intervention because the designer ought to comprehend the contemporary design sensibilities, maneuver the design's rendering style, design palette and by doing so, he or she directly meddles with the livelihoods and artisan skills of the craftsman [37]. So, sustainability is rather the outcome of a responsible innovating design culture that preserves the artisan skills of weaving craftsmen as well as improve their socio-economic conditions [37]. The innovative design culture aims at trans-evolving the visual design palette of handloom sarees that are subtly curated with a rearranged pattern layout combining new non-traditional imagery and new visual expressions yet retaining its distinctive design repertoire and handloom texture. Similar rationale was also sensed by the late Kamala Devi Chattopadyay, the Indian craft, and social reform movements stalwart who advocated self-sufficiency and aesthetic fulfilment for handloom weavers during the swadeshi movement [38].

Though the craftsmen handling the looms or otherwise identified as master weavers demonstrate creative skills in transforming the intricate design on the weave form, they lag on the knowledge grounds of contemporary design sensibilities wrought by lifestyle changes [39]. So, the common strategies applied by designers and policy makers in the name of intervention deals with either improvising a technique to improve the rate of production yet retain the quality and craftmanship or enhance the visual design palette of product through infusing contemporary design sensibilities and enriching the motif styles or minimalize the level of artisanship involved in the handloom designs for socio-economic concerns or tweak the handloom designs to produce a fusion look.

Three factors attributed for sustainable practices of businesses in capitalist economies are financial profitability, social responsibility, and environmental impacts [40]. Premised on these factors and inspired by Andhra Pradesh and Telangana's co-operation movement, collaborative ideas were mooted to protect economic autonomy and social autonomy of artisans by keeping them in charge of production instead of using them as wage-based laborers in mass scale production [41]. The market sector of handloom products is broadly classified as 1. Regional market sector that sells old ritual products and 2. Global cultural identity product market [37]. NGO institutions made the early attempts to define the relationships between design, technology to improve artisan production, and position the handloom products in domestic retail sector as well as Global cultural identity product market [37].

Further NGOs also donned the roles of providing market related services for design and technology assistance to improve handloom artisan livelihoods [42]. The NGO's are credited for introducing the niche handloom market space to the craftsmen in terms of visualizing social gains and financial gains, a fact acknowledged by none other than Ashoke Chatterjee, past president of the Crafts Council of India, and ex-director of NID [43]. This acknowledged initiative of NGO's foresaw the employment of designers for sharing design knowledge and developing innovative handloom designs. But later when the foreign funds dwindled in 2010, new market policies and directives fostered collaboration of e-tailers and craftsmen with the help of designers [44]. Thus, a progressive era of design for consumer led by designer activism strategies took shape. As a result, a greater number of designers collaborated with artisans taking the handloom products to Indian fashion and Global fashion as well.

4.1   Designer Intervention Strategies in Jamdani

The case of Jamdani continues to inspire the design innovations brought forth by the integration of weaver's tacit knowledge of interweaving or interlacing techniques and the designer's knowledge of engineering the rendering style to expand the traditional design palette [37]. Especially in Jamdani, where the desired design replica on graph sheet is placed under the warped threads supported by a tandem arrangement of the discontinuous weft threads of desired colors on the top of warp threads, the weaver's tacit knowledge and craft of engaging the required weft threads come into play along the process of creating the successful design.

Contemporary Jamdani combines traditional technique of interweaving with computer aided designs for producing complex design patterns extending all over the saree [37]. Jamdani owes the flexibility of producing a floral motif by interlacing one warp and one pick or grouping several warps and several picks whose numbers can range from two to four of different permutations and combinations. Hence the different permutations and combinations of the floral motif rendering is visualized on a computer aided design platform before finalizing the motif shape. Each pixel in a graph corresponds to one warp and one weft. Working with such groups of warp and weft reduces the labor involved in the production of Jamdani work. Reduced labor directly translates to cheaper saree price. To produce the Jamdani pattern, the weaver plans the lifting order of warp yarns and arranges them on the loom with the help of loom device: Jalari. The weaver follows the graph sheet to set up the Jalari. Now a days, as the graph sheet design pattern is visualized directly on the computer aided design platform and the weaver has it printed, it saves around half the time that goes into the preparation of the graph sheet [37].

Young designer Gaurang shah's work revolves around introducing new blended fabrics, colors, textures, and yarns to the erstwhile Jamdani with a purport to expand its design palette and introduce new style variants in terms of pattern arrangement and rendering. Hence, the prime role of Gaurang shah is to provide the contemporary design inputs regarding the trend information pertaining to motif style, design imagery, color ways, texture to the craftsmen besides understanding the techniques, and craft deployed by the weavers [45]. For successful translation of design details, it requires a great deal of knowledge about the variable options for the extra weft threads that make between 100–300 different types and the ornamental patterns made from them in desired size [46]. And in Jamdani, the art of transferring the design on paper to the fabric lies with every weaver's artisan skills rather than relying on a jacquard to produce the design. Hence the twin efforts of Gaurang shah and the weaver's artisanship forge continuous design exploration propelling the Jamdani ornamental work to incorporate any intricate design according to the new globe-trotting young Indian preferences and reap the economic rewards for the weaving craftsmen. One such experimentation on cotton muslin using silk, gold, and silver coupled with both traditional and non-traditional ornamental figures expanded the design palette of Jamdani sarees by adding textures and visual expressions. The core elements of Jamdani weave are the pattern, interlace or interweave, texture, and the yarns. Almost any type of pattern nevertheless its intricacy is transferred on the ground weave by the judicious choice of interlace or interweave type. The interlace or interweave ranges from twill weft floats to plain weave interlacement. The flexibility to choose the type of interlacement with reference to the intricacy of the pattern on ground weave construction makes it versatile to incorporate any type of motif edge finishes: clear edge, wavy or feathery edge, or Holbein stitch like stepped edge. These sorts of rendering style manipulation on different texture and yarn counts which are anywhere between 80 and 300 s count, provides immense scope for design exploration. Moreover, the interlaced motifs and ornamental figures are incorporated using discontinuous extra weft threads that could be of any color and type among the variable options. Thus, the color palette of the contemporary Jamdani sarees is a breakaway from the traditional varieties. They are classified into three major types as follows. 1. White colored extra weft ornamentation on natural colored (unbleached) cotton grounds or cool pastel color tones or dark ground weaves, 2. Mix and match of white and colored extra weft ornamentation work on any of the ground backgrounds of desired contrast, and 3. Zari and colored extra weft ornamentation work on any appropriate ground colors [20].

Among the pattern experiments on the Jamdani design palette, Gaurang shah's aspirational new pattern is the improvisation of traditional Maharashtrian bangle patterns comprising of four birds in each bangle motif [47]. Another pattern exploration by Designer Rahul Mishra yielded a dual hued Jamdani jackets and Jamdani sarees, where the white colored ornamental work was arranged on ground hues of indigo and cobalt blue [48]. The popular 'Tree of life' theme rendered using prints was given a new style of rendition using Jamdani zari extra weft ornamental work on khadi and silk blended ground weaves by Gaurang shah. One of the exemplary works of artisanship in the pattern explorations and representative imagery is attributed to the rendering of the Raja Ravi Varma's painting on the pallu portion of the saree. The Raja Ravi Varma's painting of Goddess Saraswathi wearing an Uppada Jamdani saree with gold zari and paisley pattern is history, as today in the contemporized version, the entire painting visual has been interwoven using Jamdani technique on the pallu portion of the saree [49]. The innovative portrayal of painting visual was created through the Gaurang shah project khadi to replicate Raja Ravi Varma's paintings on the pallu of saree using Jamdani technique with the help of Srikakulam khadi Jamdani weavers. The project took the toll of identification and development of about 600 color shades followed by dyeing over 200 kgs of yarn as preparatory work for weaving. For identifying the color shades the paintings were scaled up to 40 inches in size. A yarn count of 150 s Ne was chosen to replicate the paintings. As the colors ranged from pastel hues to luminescent hues, replicating them on 150 s count explicates the quantum of planning and artisanship. This project khadi was organized to commemorate Mahatma Gandhi's birth anniversary and Raja Ravi Varma's death anniversary [49].

Hence the versatility of Jamdani extra weft ornamental technique to render painting visuals and intricate patterns in desired rendering styles is well proven beyond any doubt. Further the material substitution in the forms of khadi, Tussar, and Muga over silk yielded new textures with contemporary design sensibilities [47].

These design innovations have not only contemporized the designs but also enriched the artisan skills of the weaving craftsmen as they are challenged to incorporate new design imagery and intricate patterns using the same extra weft interlace or interweave technique (Figs. 5 and 6).

Fig. 5 Raja Ravi Varma's painting of Goddess Saraswathi wearing an Uppada Jamdani saree. *Source* (https://commons.wikimedia.org/wiki/File:Saraswati.jpg)

Fig. 6 Contemporary Jamdani showing the entire painting visual woven on the pallu of saree. *Source* (https://www.gaurang.co)

4.2 Design Intervention in Kanjivaram Saree

One of the weakness spotted during craft cluster study on traditional Kanjivaram is lack of expansive knowledge about the motif types among the weavers [50]. Usually, a Kanjivaram style consists of colored body with gold zari interlaced border. The colored body background might also include texture patterns, motifs, and ornamental figures according to the saree design philosophy. New design variants on the likes of reverse design that is gold zari interlaced body work alongside colored border with ornamental rendering have also been explored. The core elements of Kanjivaram saree are represented by the embossed surface of rendered motifs and figures, heavy texture, and the color schemes used for rendering patterns.

In place of the traditional 2 ply silk warp and 4 ply silk weft, khadi yarns and linen yarn have been experimented by the designer Radharaman of Advaya Brand. His design philosophy "the point of design is to differentiate" is evidently seen in the novel organza, khadi, and linen blended Kanjivarams [51]. The iterative experimentation and prototyping with different weft material and new warp yarns takes an abstract path with the evocation of producing the certain property of drape and handle properties besides the exploration of colors and motif imagery [51]. These material substitutions provided alternative drape and handle deemed fit for modern day end user occasions of the kinds of casual wear and part wear. This texture friendly exploration gained momentum when the colorways and motif imagery are visualized in different ways quickly with the help of digital platforms rather than the unusual tedious process of experimenting with warp and weft. Experimenting with warp and weft involves understanding the texture and bending properties, vis a vis its compatibility to weave by hand in conventional handlooms. 3D design software with the provision to modulate fabric behavior by inputting mechanical properties served the purpose and efforts undertaken to convince the weaver who is used to weaving only the traditional silk yarn. Sometimes showing a visual prototype is required to convince the weaver. Organza Kanjivaram is one of a kind sheer weave and light in weight contrary to the heavy silk aka 'murukku pattu' Kanjivaram varieties [52]. On these organza Kanjivarams, one can see the discontinuous rendering of motifs like the Banarasi kadwa technique. These texture-based material substitutions added on a new dimension of muted tones in place of the traditional iridescent texture (Fig. 7).

Fig. 7 Kanjivaram saree showing embossed zari effect. *Source* (https://commons.wikimedia.org/wiki/File:A_silk_saree_loom_in_Kumbakonam,_Tamil_Nadu.jpg)

Brands like 'Kanakavalli' present curated versions of authentic Kanjivarams with renewed fine ornamental rendering of figures and motifs on dual tone backgrounds. The dual tone backgrounds are referred to as shot colors attributed to the translucent overlay effect of one color on the other. The warp and weft are dyed in different hues to create this overlay color effect. This fine rendering transformation is brought forth by the migration of hand drawn motifs on graph sheets to computerized versions made on photoshop, weaving CAD software, and MS paint platforms. The traditional Jala or Adai technique are still prevalent, but visualizing designs on digital platforms with the help of designers has become a regular practice owing to the continuous interaction with the weavers. The reason cited by buyers for continual preference of Jala or Adai weaving over jacquard versions is attributed to the embossed effect of extra warp and extra weft designs obtained through it [53]. Lately subtle overture effects amidst the regular composition of motifs, and ornamental figures have provided a breakthrough in the number of weft colors used in weaving extended up to four colorways (Fig. 8).

Majority of the commercial Kanjivaram brands likes of Nalli silks, Bharani, Chennai silks, etc. have expanded the color palette of the traditional Kanjivaram design palette by roping in fashion hues like koral, royal blue, and mint green to name a few of the whole lot. Designers like Santosh Parekh even classified the color palettes of Kanjivaram produced by his design house; Tulsi silks as Pastels, Musicals, Animals, Brights, and Golds [54]. Each category has a distinctive color palette inspired by a theme. These brands also offer extended services like producing on demand 'One of a kind exquisite designer saree' for bridal ceremonies. The latest addition to the Kanjivaram design palette is the Kanjivaram lehengas launched by Soumya Nandivada. The new attiring style of Kanjivaram also includes contrast colored silk blouses with fusion embroidery [55].

Hence the heterogeneity of Kanjivaram design lies in expanding its design palette to especially the sheer, translucent properties of the fabric appearance as well as exploring the motif imagery, and rendering styles. This sort of change in visual appearance brought forth changing the yarn texture and type makes it a sought-after design for parties and other occasions. Explorations on the overtures and the curated embossed surface effect of the rendered ornamental figures and motifs has uplifted the style of Kanjivaram sarees in tune with the deconstructed graphic design sensibilities. Further the new color schemes have opened the visual expression of Kanjivaram sarees in new color contrasts.

The exquisite Kanjivaram design palette with all over ornamental figures, we see today, came into existence in 1980s with the introduction of jacquard mechanism [56]. These jacquard produced Kanjivaram saree co-exist with curated versions of Kanjivaram sarees produced using the traditional Jala technique or Adai technique. Ever since the Kanjivaram was patronized by the late Bharatanatyam dancer, Rukmini Arundale Devi, its design palette has undergone many transformations. She combined border design from one saree, pallu design from another with distribution of traditional motifs across the body, an eclectic combination that still inspires today's contemporary design explorations [57]. Another way of adding exquisiteness to Kanjivaram design palette includes experimentations like using gold plated zari

Fig. 8 A modified Kanjivaram saree with overture effect. *Source* (Doddampalayam saree weaving cluster)

in both warp and weft. The iridescent look is the prominent feature of Kanjivaram design palette. Alternatively, the designers have forayed into pastel colors also.

Another versatile aspect of Kanjivaram is its ability to accommodate intricate rendering styles of any nature. In one of the efforts, inspired by Victor Vasarely's optical patterns, a leading retailer RMKV of Tamil Nadu produced a patented design using multiple colors [58]. This innovation is attributed to the replacement of petni technique by tie and dye technique that renders the possibility of incorporating kalamkari style sharp motif shapes on Kanjivaram besides cutting the human labor part. Another notable intervention that helped them to produce innovative designs is the pneumatic assembly that uses compressed air to place and displace the jacquard mechanism on top of loom without human efforts [58].

Unlike Banarasi sarees in regular color palette of red, blue, and pink, the Kanjivaram offers an extended color palette including fashion colors ably supported by commercial color labs [59]. The Kanjivaram reflects the light owing to their iridescent nature and shimmering surface, an intrinsic property which produces ombre effect amidst the gallery lighting [59]. These visual effects are the signature preference of the current day Youngsters thriving on selfie culture, WhatsApp, and Instagram. Designer Neeta Lulla infused western motifs like polka dots on lighter Kanjivarams to suit the younger lot across modern outfits: Pant suits and lehengas. Meanwhile designers like Gaurang Shah further extended the Kanjivaram design palette with bigger border design and fresh neon fashion colors [60].

Kanjivaram saree design palette has imbibed both engineered designs through jacquard mechanism and hand curated embossed effects that rely on the artisanship of the weaving craftsmen who use the Jala technique. Both the design rendering techniques use the same design repertoire elements in different ways unique to them to yield varied visual expressions. Thus, these design explorations have added on cultural diversity and heterogeneity to the Kanjivaram saree design palette.

4.2.1 Technological Innovations

Another instance is the enforcement of child labor act that nearly eclipsed the korvai technique used in Kanjivaram saree. The child labor act banned the children below of the age of 14 to participate in the weaving process against the usual practice for the weaver's children to take part in the weaving process alongside their parents. However, a late invention, SPS korvai sleigh developed by an artisan revived the korvai technique as the weaver himself can carry out the knotting technique without the help of cheap labor [61]. And tie and dye method has replaced the traditional petni method. Additionally, the visualization of designs in digital platforms have further cut down the arduous task of preparing the designs on graph sheets prior to weaving.

The tie and dye technique is borrowed from Dharmavaram known by the name Reku—tie and dye technique. In Reku technique, total length of warp required to make 4 sarees is taken and folded in plies, with each folded ply length equaling the saree length. Each folded length of warp represents the total saree length inclusive of body warp and pallu warp. A demarcating line between body warp and pallu warp is marked on the folded sheets of ply. The pallu part is tied and covered with polythene sheet. First the body length of folded sheet is dyed in required body warp color. After dying the body warp, it is tied and covered, meanwhile the pallu warp is untied, uncovered, and dyed in required color. Thus, the body warp and pallu warp are dyed alternatively one after another a couple of times. At the demarcating line which indicates the body warp section boundary and pallu warp section boundary the colors overlap for a length of 1 ½ in. Now the dyed body warp ends, and border ends are gaited to the older ends in the beam. The pallu part is woven first incorporating all the extra weft designs. As we reach the cross over section that demarcates body warp length and pallu warp length, an assessment about the length covering the overlapping colors is made and necessary adjustments are planned. The adjustments are generally zari weft thread woven in weft faced twill or sateen to cover the excess and irregular color overlapping regions at cross segment. For weaving the body warp and border warp, the entire number of warp ends are divided into two layers with even numbered ends forming one layer of warp and odd number of ends forming another layer of warp. This process of division is carried out by splitting the odd numbered ends and even numbered ends by placing a lease rod between the heald frame and warp beam. These two layers are referred to as "Rekus" in Telugu. Reku means layers. Now the layered warp ends are loosed down, and two lease rods are inserted one after another across the top warp layer, with the top warp layer making alternating half wraps around the lease rods. This task makes the top warp layer to be held tight. This wrapping junction covers a length of 2–4″ and acts as a firming point for the top warp layer. Meanwhile the bottom layer is loose for the same length of 2–4″. Now the loose bottom layer is lifted up to the cloth fell and a lease rod is inserted before the reed to arrest the sagging behavior of bottom layer. Later the lease rod is removed as the cloth roll is rotated to remove the sagginess in the bottom layer. Hence both the layers are tight now. This pulling back recovers the 2–4″ length of the bottom layer which has overlapping colors. Meanwhile the lease rods inserted in the top layer of warp is removed and the act of lease rod insertion at the cloth fell before reed and rolling the saggy top layer is carried out. But here the lease rod is tied firmly, and the fabric weaving is continued for 5″-6″ of length. After weaving around 5″–6″ of fabric the tied lease rod is removed. Further to this, the saggy part containing the overlapping colors form loops which are cut leaving 1″ above, so the excess crossover color part is adjusted by this Reku technique. The advantage of Reku technique is it needs near about an hour to complete the work whereas kondi and petni work requires one full day. So almost most of the weavers now practice Reku technique for medium cost range sarees whereas if the product requirement, cost affordability goes hand in hand only then the weavers practice kondi or petni technique.

4.3   Design Intervention in Ikat

Ikat designs remains the most coveted among the handloom products as the design by nature are still produced using human efforts and made by their hands. The quantum of Ikat design work relies upon the practice and skill of craftsmen to dye the yarns and replicate them, though today the designs are visualized in photoshop for clarity and understanding. The core of Ikat design palette comprise the elements of color, pattern, and texture.

One such design intervention on the element color was pioneered by Maharaja Sayajirao University faculty owing to the paucity of natural resources, lack of contemporary colors, and cumbersome process of extracting natural dyes. The project embarked upon training the weaving craftsmen to carry out the tie and dye work with the help of industrial reactive dyes. The project investigators initially transferred the Ikat designs on paper to Adobe Photoshop populated with pixels [62]. The chosen photoshop work area was 60 X 60 pixel square, though on the given page layout it is possible to choose any given number of pixels in accordance with the complexity of the design. Later the chosen designs were dyed using reactive dyes instead of the traditional dyes and the Ikat fabric was transferred to garments. The study embarked upon the choice of reactive dyes for Ikat made of cotton yarns is a game changer as it lends a fashion forward color palette yet retains ingenious ornamental rendering style. Prior to the production, the artisans were trained on the dyeing of the warp and weft in reactive dyes [62].

Among the textural explorations, Designer Ritu Kumar, associated with neopatna cluster craftsmen to produce a collection of light weight Ikat fabrics that were used for Casual apparel tops [63]. Though the same craftsmen had the potential to weave the ten avatars of Lord Vishnu in figure shapes and the verses of Gita Govinda on the fabric, it did not have enough takers owing to its heavy weight except the temple requirements [63]. Designer Anupama through one of the unique design contemporizing maneuvers, converted 25-year-old Ikat sarees into a western gown with shirt collar [64]. Ikat from three different regions—Odisha, Gujarat, and Andhra Pradesh—were combined in an eclectic way to produce Boho looks western styled gown [64]. The intention is to create a new style that is wearable in the contemporary occasions using the already available aesthetic fabric, a kind of upcycling task. This was not a first time; three varied rendered style patterns of different scale were framed together as one single visual palette rhyming with the visual effect "Unity in diversity". Eclectic design explorations have always been a part of design intervention strategies premising from the fusion concepts.

Earlier Designer Madhu Jain had also combined three different forms of Ikat pertaining to Indonesia, Uzbekistan, and India to bring out fusion Ikat patterns [65]. Additionally, she also introduced material variations in the form of bamboo yarns and silk yarns for Ikat designs adding new dimension of texture. The most flamboyant of the design innovations involving texture in Ikat was brought forth through designer Sravan Gajam by combining the handwoven warp denim and Ikat technique [66]. Abstract motifs alongside traditional ones were ushered to define a novel indigo hued denim Ikat across all denim specific textures like ripped, distressed, and stretch marks.

These design innovations using the elements of color and texture has added new dimensions with contemporary design sensibilities. The introduction of reactive dyes for Ikat has added flexibility to use multiple dye baths thus yielding more vibrant color schemes. The extension of Ikat design palette has not only been enhanced its contemporary design value but has also enriched the weaving craftsmen knowledge of using the dyes and new materials. Since the Ikat design patterns are pre-formed on the surface of yarn which gets transformed upon weaving, it largely relies on the skill and knowledge of the weaving craftsmen who balances the yarn hue and shades according to the pattern. And to compound the case, there is still no machine or automation available to process the final design plan into broken down forms ready made with yarn dyeing pattern information and quantify the type, amount of dye required [67]. So, the skill and artisanship of weaving craftsmen assumes utmost significance for producing the design plans pertaining to order of yarn dyeing.

4.3.1 Technological Developments in Ikat Process

A 4 spindle, domestic hank to bobbin winding machine was invented in (2011) to lower the amount of labor in winding operation. The traditional winding method uses a hand rotated charka with which only 100–120 hanks of 840 yards are possible in a day. The hank winding machine is operated with a miniature motor of 60 w that could be fueled by 12 V batteries as well or can be operated with pedal motion by legs. These hanks are grouped for tying and dyeing. A subgroup of yarns comprises of two and four yarns in Odisha Ikat whereas it is eight yarns for Pochampally Ikat. Hence it is an affordable choice for the handloom weaver [67].

The most tedious process in Ikat is tying and dyeing process. The master weaver interprets the number of groups of yarns required and pattern of knotting to achieve the design repeat. A great deal of preparatory calculation goes into the preparation of yarn groups for dyeing that is different in Odisha, Gujarat, and Andhra Pradesh. In Andhra Pradesh, instead of tying and grouping the yarns in triangular Ikat frames, Asu process a recent innovation by Sri Mallesham consists of an electronic device and automatic grouping device. This machine provides the flexibility to adjust the number of threads in each peg. Whereas the traditional technique of winding warp yarns manually on pegs for tying and dyeing takes about 5 h and strenuous manual efforts to complete the task meant for one saree. However, with the innovative yarn winding machine the winding job time has been cut down and it is possible to wind for six sarees in a single day [68].

The Automatic Asu winding machine is suited for only Andhra Pradesh style of grouping the yarns in triangular frames that constitutes a central peg and forty other peripheral pegs. For rectangular frame grouping of yarns used by the weavers of Gujarat and Odisha, a separate modified warping drum assembled with a worm gear and an electronic device for sorting and arranging the groups of yarns has been developed by Nuapatna Weaving cluster through GOO-UNIDO project [67]. The invention is on the verge of commercialization. Since the worm gear mechanically controls the traverse distance upon inputs from the electronic device, the manual efforts required to sort, and arrange the groups has been reduced drastically.

4.4 Impact of Designer Intervention Strategies on Weaving Craftsmen Livelihoods

Scientist S. A. Hipparagi of CSTRI, confirmed that sarees worth Rs. 25,000/- and above with complex designs fetches better returns for the weaving craftsmen compared to the moderate designs sold at lower prices. And the weaving craftsmen working with designers and recognized clusters have been engaged continuously which provides them uninterrupted livelihoods. But the weavers who do job work for buyers and local middlemen without any knowledge of the saree's identity, like the case of weavers of Y.N. Hoskote are the ones who are affected by the economic travails such as silk yarn price hike and materials shortage [69]. And weaving craftsmen those who are self-employed rely on the local seasonal orders and direct orders have been the most affected [70]. About 40,000 weavers of the non-profit organization, Banka silks associated with Indian federation for fashion development (IFFD) which organizes India runway week can sell their produce directly to leading Indian fashion designers at fair prices [71]. Leading Indian fashion designers like Rajesh Pratap Singh, Rahul Mishra, Abraham & Thakore and Samant Chauhan assigned to handloom clusters for product development are entrusted with the prime task of upgrading the weaving craftsmen skills for the twenty-first century buyer [72]. The weavers enrolled in this scheme demonstrate new confidence levels to take on the market and paid well. Government of India has also set up design resource centers through design school NIFT with the sole purpose of equipping the weaving craftsmen to co-create design excellence in handloom sector targeting global fashion exports [73]. Thus, the weaving craftsmen associated with design centers or leading fashion designers or design schools stand to benefit both in the form of enriched artisan skills and continuous employment with fair remuneration.

5   Conclusion

The infusion of non-traditional imagery, new visual expressions, and their combination with traditional elements of cultural identity has added contemporary design sensibilities to the handloom design palette making it more diverse and heterogenic. In other words, the distinctive cultural identity of handloom design palette is integrated with the cyclic fashion trends by the designers to suit the twenty-first century young Indian consumers. And the technological innovations in the forms of machinery, equipment, and methods have reigned in productivity and quality amidst retaining the distinctive attribute of handloom processed fabric texture. With reference to the available literature about the artisan skills of rendering intricate designs as imagined by the designers and technocrats, it is concluded that the design palette and the repertoire of the elements are capable of being deconstructed and shaped according to the twenty-first century buyer's lifestyle needs. The design intervention strategies have also served to enhance the weaver's skill set of rendering the motifs and patterns by experimenting with new visual imagery and fusion concepts. The weaving craftsmen working with the designers, big designer brands, and Dastkari committees are not only able to interpret the new age buyer preferences through the ubiquitous design experimentations but also benefitted in terms of better remuneration and continuous employment. While weavers those who are self-employed or work on contract basis with local merchants seem to be affected by the market factors. The handloom designs discussed in this chapter of the capability to produce customized premium designs with comparatively less dependence on electricity across rural households of India holds promise and potential that shall fetch better rural employment, reverse rural to urban migration, and create new sustainable business models. However, the findings cannot be generalized to other handloom saree designs and non-apparel handloom products owing to the diversity of the design repertoire. Overall, the study helped to conclude the scope and potential of design intervention strategies for achieving cultural and economic sustainability but could not ascertain the environment impact which if carried in near future shall provide a more detailed idea for further progressive developments along the course of environmental sustainability.

References

1. Kaushik KV, Khanna A, Sah S (2019) Indian Handloom industry–Position paper. Available via DIALOG. http://www.ficciflo.com/wp-content/uploads/2019/03/Indian-Handloom-Industry-Final. Accessed 9 December 2020
2. Fourth All India Handloom census 2019–2020 (2019) Ministry of textiles, government of India, New Delhi. Available via DIALOG. http://handlooms.nic.in/writereaddata/3736. Accessed 30 December 2020
3. Indian handloom industry: Potential and prospects (2018) Export Import bank of India, New Delhi. Available via DIALOG. https://www.eximbankindia.in/Assets/Dynamic/PDF/Public ation-Resources/ResearchPapers/102file. Accessed 2 January 2021
4. Raniwala P (2019) What goes into the making of a pure gold zari sari? Available via DIALOG. https://www.vogue.in/content/how-is-pure-gold-silk-saree-swati-sunaina. Accessed 12 November 2020
5. Kalyani A, Rohitha V, Bharathi M (2017) An analytical study on issues of handloom industry in undivided State of Andhra Pradesh. Int J Innovat Res Explor 4(6):1–10
6. Zutshi V (2019) August 7 was National handloom day–but the handloom sector still faces huge challenges. Available via DIALOG. https://www.thehindu.com/society/august-7-was-national-handloom-day-but-the-handloom-sector-still-faces-huge-challenges. Accessed 27 December 2020
7. Two-week social media campaign under hashtag #Vocal4Handmade launched to promote handloom products (2020), Press information bureau, New Delhi. https://pib.gov.in/PressRelease Page.aspx?PRID=1644093. Accessed 10 December 2020
8. Gogna N (2020) Indian handloom industry: Misery spells 'Opportunity'! Available via DIALOG. https://www.tpci.in/indiabusinesstrade/blogs/indian-handloom-industry-misery-spells-opportunity. Accessed 29 December 2020
9. Mathur V (2020) Design inspiration: motifs of the beautiful Kanjivarams of India. Available at:https://uxdesign.cc/motifs-of-the-beautiful-kanjivarams-of-india. Accessed 11 December 2020
10. Kanakavalli Varnsutra (2018) Kancheepuram–The Indian Saree Journal blogpost. Available via DIALOG. https://kanakavalli.com/blogs/kanakavalli-journal/varna-sutra-haridhra-kanjiv arams-colours-of-the-sun. Accessed 12 December 2020
11. Panneerselvam RG (2014) Petni. Kondi and Reku, traditional techniques of weaving handloom silk sarees. Indian J Tradit Knowl 13(4):778–787
12. Raniwala P (2019) What makes Gujrati Patola Sari a priceless heirloom? Available via DIALOG. https://www.vogue.in/content/gujarati-patola-sarees-history-significance-weaving-process. Accessed 14 December 2020

13. Yashita (2019) Best and trending patola sarees that every woman needs to own! Available via DIALOG. https://stylesatlife.com/articles/Patola-sarees. Accessed 11 December 2020
14. Pochampally ikat patola silk sarees (2020) Available via DIALOG. https://www.pochampal lysarees.com/product-category/pochampally-ikkat-tradtional-sarees. Accessed 11 December 2020
15. Pochampally sari (2020) Available via DIALOG. https://en.wikipedia.org/wiki/Pochampally_ sari. Accessed 12 December 2020
16. Pochampally ikat sarees: the history of india's famous double ikat weave (2017) Craftsvilla. Available via DIALOG. https://www.craftsvilla.com/blog/pochampally-ikat-sarees-the-his tory-of-indias-famous-double-ikat-weave. Accessed 12 December 2020
17. Mohanty BC, Krishna K (1974) Ikat fabrics of Orissa and Andhra Pradesh. Calico Museum of Textiles, vol 1. Ahmedabad, India, p 15
18. The Sambalpuri ikat of Odisha: History, symbolism, and contemporary trends (2017). In: Visual and material arts, Sahapedia.Org. Available via DIALOG. https://www.sahapedia.org/the-sam balpuri-ikat-of-odisha-history-symbolism-and-contemporary-trends. Accessed 14 December 2020
19. The handloom weaving process (2017) Available via DIALOG. http://www.sarasu.in/The-Han dloom-Weaving-Process-pid>. Accessed 12 December 2020
20. Das S (2014) Case study on the textile revivalist and entrepreneur designer "Gaurang shah's approach in contemporizing jamdani (extra weft insertion technique)". Dissertation, NIFT, New Delhi
21. Jamdani: weaving history (2011). Available via DIALOG. https://www.dawn.com/news/630 713/jamdani-weaving-history. Accessed 11 December 2020
22. Naidu TA (2020) Of Uppada and weaving skills of girls. Available via DIALOG. https:// www.thehindu.com/news/national/andhra-pradesh/of-uppada-and-weaving-skills-of-girls. Accessed 30 December 2020
23. Sarode G (2009) Available via DIALOG. https://sarode1.wordpress.com/2009/05/07/ghansh yam-sarode. Accessed 30 December 2020
24. Uppada jamdani sarees (2020) Available via DIALOG. https://www.utsavpedia.com/attires/ uppada-jamdani-sarees. Accessed 11 December 2020
25. Jamdani Silk (2020) Available via DIALOG. https://www.utsavpedia.com/motifs-embroider ies/jamdani-silk-a-tradtional-weave/. Accessed 11 December 2020
26. Menon M (2020) Bhimbatka: India's oldest art gallery. Available via DIALOG. https://www.liv ehistoryindia.com/amazing-india/2020/04/01/bhimbetka-India's-oldest-Art-gallery. Accessed 14 December 2020
27. Mishra VN (1981) The prehistoric rock art of Bhimbetka. Central India. J Nat Cent Perform Arts 10 (1):1–16
28. Taylor SP (2016) Aesthetics of sovereignty: the poetic and material worlds of medieval jainism. Dissertation. University of Pennsylvania, USA
29. Bayly CA (1986) The origins of Swadeshi (Home Industry): cloth and indian society 1700– 1930. In: Appadurai A (ed) The social life of things: Commodities in cultural perspective. Cambridge University Press, Cambridge, pp 285–322
30. Dutta D, Saxena TG (2009) International brands: India entry strategies. Available via DIALOG. http://www1.udel.edu/fiber/issue4/world/internationalbrands. Accessed 3 January 2021
31. Shroff Y (2017) The fascinating evolution of Indian Women's Fashion over the years. Available via DIALOG. https://yourstory.com/2017/03/fascinating-evolution-indian-womens-fashion-years>. Accessed 30 December 2020
32. Ranavaade VP, Karolia A (2016) National fashion identities of the Indian sari. Int J Tex Fash Technol 6(4):15–18
33. Agarwal P (2020) These 5 companies' ethnic wear ranges are winning the market despite competition from designer brands. Available via DIALOG. https://yourstory.com/smbstory/eth nic-wear-traditional-silk-sarees-lehenga-suits-designer-brands. Accessed 30 December 2020
34. Ranavaade VP, Karolia A (2017) The study of the Indian fashion system with a special emphasis on women's everyday wear. Int J Tex Fash Technol 7(2):27–44

35. Ranavaade VP (2017) A semiotic study of the Indian Sari. PhD Dissertation. The Maharaja Sayajirao university of Baroda, Gujarat, India
36. Telier A, Binder T, De Michelis G, Ehn P, Jacucci G, Wagner I (2011) Design things. The MIT Press, Cambridge MA
37. Mamidipudi A (2016) Towards a theory of innovation in handloom weaving in India. PhD Dissertation. The Faculty of Arts and Social Sciences, Maastricht University, Netherlands
38. Priya M (2018) Kamala devi Chattopadyay - The torchbearer of Indian crafts. Available via DIALOG. https://www.sarangithestore.com/blogs/sarangi-journal/kamaladevi-chattopad hyay-the-torchbearer-of-indian-crafts. Accessed 13 December 2020
39. Liebl M, Roy T (2004) Handmade in India: Traditional craft skills in a changing world. In: Finger MJ and Schuler P (eds) Poor people's knowledge: promoting intellectual property in developing countries. World Bank and Oxford University Press, Washington, pp 53–72
40. Elkington J (1998) Cannibals with forks. New society publishers, Gabriola Island
41. McGowan A (2009) Crafting the nation in colonial India. Palgrave Macmillan, Newyork, p 140
42. Niranjana S, Vinayan S (2001) Growth and prospects of the handloom industry. Planning Commission, New Delhi: Government of India
43. Chatterjee A (2007) Artisan enterprise: development, cultural property and the global market. Int J Intang Cult Herit J Arch, p 3. Available via DIALOG. https://globalinch.org/article/ Artisan enterprise: Development, Cultural Property and the Global market. Accessed 2 January 2020
44. Can our future be handmade? (2019) International journal of intangible cultural heritage, 2. ISSN: 2581–9410. Available via DIALOG. https://globalinch.org/article/ Can our future be handmade. Accessed 6 January 2020
45. Fashion shows and Trade fairs (2013) In: RTW Webzine 2:66. Available via DIALOG. https://issuu.com/rtwmag/docs/rtw_webzine__2. Accessed 3 January 2021
46. Gandhi S (2019) Jamdani saris: history, evolution and how they are woven. Available via DIALOG. https://www.vogue.in/content/jamdani-saree-history-origin-technique-indian-handloom. Accessed 2 January 2021
47. Sangeetha Devi Dundoo (2020) CCT spaces opens with an exhibition of jamdani weave by Gaurang shah. Available via DIALOG. https://www.thehindu.com/life-and-style-fashion/cra fts-council-of-telengana-opens-cct-spaces-in-hyderabad. Accessed 21 December 2020
48. Rahul Mishra 2014 spring summer collection (2020) Available via DIALOG. https://www.not justalabel.com/rahul-misra. Accessed 11 November 2020
49. Team Asianet Newsable (2020) Fashion designer Gaurang shah talks about his idea behind Raja Ravi Varma project. Available via DIALOG. https://newsable.asianetnews.com/lif estyle/fashion-designer-gaurang-shah-talks-about-his-idea-behind-raja-ravi-varma-project. Accessed 27 December 2020
50. Pooja Ostwal, Pooja VV, Kuppuram N, Prasad R, Sharma S (2019) Craft cluster-kanchipuram. Final year project, Department of fashion communication, NIFT, Bengaluru
51. Shalini Shah (2019) Evolution of Kanjeevaram saris from organza to linen blends. Available via DIALOG. https://www.vogue.in/content/kanjeevaram-sarees-evolution-from-silk-to-khadi-organza-advaya-angadi-galleria-interview. Accessed 14 December 2020
52. Southindiafashion (2018) Beautiful kanchi Organza saree looks of celebrities. Available via DIALOG. https://www.southindiafashion.com/2018/03/kanchi-organza-sarees-looks-of-celebrities. Accessed 01 November 2020
53. Verma R (2017) Kanjivaram weaving: the infrastructure and technology used. Available via DIALOG. https://artsandculture.google.com/exhibit/kanjivaram-weaving-the-infras tructure-and-technology-used-dastkari-haat-samiti. Accessed 04 November 2020
54. Parekh S (2016) Lakme fashion week 2016: Day 6. Available via DIALOG. https://pho togallery.indiatimes.com/fashion/indian-shows/lfw-16-day-6-santosh-parekh/articleshow/539 06436. Accessed 03 November 2020
55. Saxena A (2020) 15 Gorgeous Kanjivaram saree designs to kick start the wedding division. Available via DIALOG. https://www.weddingwire.in/wedding-tips/kanjivaram-saree/. Accessed 21 November 2020

56. Kawlra A (2014) Duplicating the local: GI and the politics of 'place' in kanchipuram. Available via DIALOG. https://www.academia.edu/7635756/Duplicating_the_local_GI_and_the_politics_of_place_in_southindia. Accessed 30 December 2020
57. Jay P (2019) The handloom: kanjivaram silk. Available via DIALOG. https://www.angadigalleria.com/handloom-kanjivaram-silk. Accessed 12 November 2020
58. Kumar P (2020) When tradition undergoes a makeover. Available via DIALOG. https://www.thehindu.com/life-and-style/fashion/when-tradition-undergoes-a-makeover/. Accessed 22 December 2020
59. Ganguly R, Singha S (2016) How kanjeevaram won the bride war over Benarasi? Available via DIALOG. https://timesofindia.indiatimes.com/life-style/fashion/buzz/How-Kanjeevaram-won-the-bride-war-over-benarasi. Accessed 21 December 2020
60. Kapur M (2014) Kanjeevaram weaves a new style. Available via DIALOG. https://www.business-standard.com/article/beyond-business/kanjeevaram-weaves-a-new-style. Accessed 22 December 2020
61. Padmanabhan G (2012) Has the Kanjeevaram saree lost its sheen? Available via DIALOG. https://www.thehindu.com/features/magazine/has-the-kanjeevaram-saree-lost-its-sheen/. Accessed 23 December 2020
62. Saiyed SS, Bhatia R (2016) Engineered Ikat Textile of Gujarat–a design intervention. In: Textile Society of America 15th Biennial Symposium Proceedings. University of Nebraska – Lincoln. Available via DIALOG. http://digitalcommons.unl.edu/tsaconf. Accessed 29 December 2020
63. Kher R (2017) Design on ikat: how the Indian fashion industry is helping revive the traditional textile technique. Available via DIALOG. https://www.firstpost.com/living/designs-on-ikat-how-the-indian-fashion-industry-is-helping-revive-the-traditional-textile-technique. Accessed 30 December 2020
64. Pasricha A (2016) Perfect Patola: a memoir. In: International textile and Apparel Association (ITAA) Conference proceedings. Iowa State university. Ohio, USA. Available via DIALOG. https://lib.dr.iastate.edu/itaa_proceedings/2018/design/21. Accessed 22 December 2020
65. Ahuja S (2017) Sustainable fashion: Veteran designer Madhu Jain innovates with bamboo-silk ikat. Available via DIALOG. https://www.hindustantimes.com/fashion-and-trends/sustainable-fashion-veteran-designer-madhu-jain-innovates-with-bamboo-silk-ikat. Accessed 22 December 2020
66. Times of India (2018) City designer gives ikat a trendy makeover. Available via DIALOG. https://timesofindia.indiatimes.com/entertainment/events/hyderabad/city-designer-gives-ikat-a-trendy-makeover. Accessed 23 December 2020
67. Behera S, Khandual A, Luximon Y (2019) An insight into the Ikat technology in India: ancient to modern era. IOSR J Polym Text Eng 6(1):28–51
68. For the Sake of Mother's Pain: Asu Yarn Winding Machine, (2008) A Dialogue on people's creativity. Exp Innov Honey Bee 19(3):4–6
69. Bhuvaneshwari S (2019) The Kanjeevaram saris, made by weavers of Karnataka's Pavagada Taluk. Available via DIALOG. https://www.thehindu.com/news/national/karnataka/the-kanjeevaram-saris-of-pavagada-taluk/. Accessed 23 December 2020
70. Singh S (2020) Will COVID-19 Worsen The Situation Of Indian Artisans & Weavers? Available via DIALOG. https://www.iknockfashion.com/will-covid-19-worsen-the-situation-of-indian-artisans-weavers. Accessed 15 November 2020
71. Banerjee A (2018) Thread of hope: weaver community in India continues to languish in obscurity. Available via DIALOG. https://www.financialexpress.com/lifestyle/thread-of-hope-weaver-community-in- india-continues-to-languish-in-obscurity. Accessed 11 November 2020
72. Sinha C (2017) Hope looms: unique government-designer partnership to rescue handloom sector. Available via DIALOG. https://www.indiatoday.in/magazine/nation/story/20170710-handloom-textile-industry-government-designer-partnership. Accessed 4 January 2020
73. Make-In-India' Program For Weavers (2020) Press information bureau, Delhi. https://pib.gov.in/PressReleaseIframePage. Accessed 30 December 2020

How Translating Between Heritage and Contemporary Fashion Can Create a Sustainable Fashion Movement

Dorothee Sarah Spehar

Abstract Sustainable fashion and the search for more balanced and less exploitative production and consumption patterns has become a fashion trend. The industry and end consumer market are currently re-discovering the values of handmade items. With practices of traditional heritage textile manufacturing, the perception and demand for artisan fashion became a popular add on to contemporary fashion designs. This article is looking into the different approaches brands and designers can have, when working with artisan communities. The findings and examples are based on expert interviews, which I conducted with the creative directors and owners of three different brands, located in India and Columbia. The value of handcraft items is shown here as a point of uniqueness, entrepreneurial business aspect and wider social responsibility. With all brands and designers having a strong bond to their local communities, they see the identity of handmade textiles as a signature to their contemporary designs. Additionally, we can emphasis the complexity of collaborating with artisan groups in a respectful way. Implementing modernisation and keeping traditions integer is a crucial pint when translating heritage skills into the future.

Keywords Heritage textiles · Artisan · Contemporary fashion · Sustainable fashion · Handmade · Communities · Slow fashion · Entrepreneurial1

1 Introduction

Sustainable fashion with all it's side topics and socio-political dynamics became one of the biggest buzzwords of the textile industry. We are facing a situation in which its becomes clear that the current setup for production and consumption are not maintainable anymore, with now consumers and industry professionals equally on the search for better ways to approach fashion. One of the many practices and focus points which arise from the trending topic, is looking back into heritage and artisan textile skills.

D. S. Spehar (✉)
DS Agency, Skalitzer Strasse 104, 10997 Berlin, Germany
e-mail: ds@dsagcy.com

© The Author(s), under exclusive license to Springer Nature Singapore Pte Ltd. 2021   251
M. Á Gardetti and S. S. Muthu (eds.), *Handloom Sustainability and Culture*,
Sustainable Textiles: Production, Processing, Manufacturing & Chemistry,
https://doi.org/10.1007/978-981-16-5967-6_11

Artisan textiles or handmade fashion fall under the perception of slow fashion, mindful consumption and the new value system of durability and uniqueness [1].

From all parts of the industry, being design, production, sourcing, or marketing, the emphasis on craft is becoming a stamp of approval when talking about sustainable fashion practices.

For artisan groups and craftsmanship communities this means a growing interest in their work and therefore business and collaboration possibilities. These opportunities are especially meaningful, when we consider some of the circumstances artisan communities are exposed to. The critical point and relevant consideration is, how to maintain cultural boundaries and heritage textiles while implementing the traditional skills into innovative designs and contemporary fashion items.

In theory, equally much as the artisans need to have the will to grow, the clients and the mainstream market needs to re-learn and re-understand the value of slowly fabricated products. This is a bigger vision for the textile industry and it is a look into a more sustainable fashion future, that proves that handmade fabrics can inspire the future of sustainability.

The following paper is an attempt to understand and spotlight several approaches to work with artisan and indigenous textile craftsman groups in India and Colombia. My own experiences as sustainability and textile compliance expert, shaped the fundamental understanding of the complex processes within fashion supply chains.

The research and outcomes are based on interviews with three different designers, who work with the intersection of contemporary fashion and heritage textile craft. I chose to interview these three designers, based on the fact that all have an international client base for their well performing fashion brands, and deliberately chose to support and implement their local artisan groups and textile aesthetics to brand identity.

2 The Value of Artisan Knowledge

2.1 Consumer and Industry Sentiment on Handmade Textiles

One of the greatest legacies that we have in the world of textiles is the knowledge, the heritage, and enormous potential that artisan communities have through their craft of handmade and handloom textiles. As a counterpart to any kind of textile mass production the concept of implementing legacy and honoring heritage is something that can be incredibly valuable when looking at more sustainable, more ethical, and more responsible ways to reinvent and restructure the fashion and textile business.

A 2018 conducted report by The Nest showed that the United States is one the forefront of countries where an overwhelming 77% of designers are motivated and interested in souring materials from artisans [2]. The tendencies and options to go forward with embedding the traditional and heritage skills more and more back in our economy are there, but how is it that we still face an value action gap when it comes to respectful and safe approaches in textile souring with artisan and handwork communities?

A current fashion system that is so much based on distance supply chains outsourcing work and exploiting labor can only benefit and learn from a way to narrate and conclude business in a slower and less capitalized manner [3]. Our habits are far away from the actual product, and its source and valuable impact. Clients and customers are conditioned to not related to the garment that they are buying because only this way the agenda of fast fashion and exchangeable items can be held up.

This is a story that has been fostered and nurtured for generations now, economical growth and the added surplus in producing and selling goods. Since the global pandemic hit countries and societies in 2020, we see shifts in the so long system of the global fashion world though. As a result of the COVID-19 crisis, 57% of the 2000 consumers part of a German and the United Kingdom-based research were to state that they 'made significant changes to their lifestyle to lessen their environmental impact' [4].

The value of the slower and old methods of hand-looming artists on textiles are something that cannot be replaced, the knowledge and heritage that is anchored in the cultural context of countries and communities. The important thing here is to understand that in order to survive, just like many other skills and industry, handmade textiles need to be newly interpreted. The world that we are living in with all of our global connections fast pushed trends and how social media changed our way to communicate, the old and the tradition can only be kept alive when adapting to the modern frameworks with authenticity and honesty. We want something understandable for us, it has to be valuable, but we also want something that is associated with innovation, urban style, and freedom of the creative mind. The proposed idea and way to introduce new opportunities would be to use these textile traditions as handlooms and hand crochets and then automatically translate them into something new, something contemporary and relatable for younger generations.

For a lot of brands and companies, the question occurs about how to maintain such a concept. How do you establish a connection with artists and communities, often rural areas, and how do you introduce change without interfering with traditions, boundaries, and cultural contexts?!

3 Contemporary Brands Working with Artisan Communities

Globally Relevant Examples of Local Craft

With the following cases of three aesthetically very different brands, I will highlight some approaches how modern fashion brands addressed and implemented the possibilities and challenges of working with artisan groups in their respective regions. The selected brands are all functioning companies, with and international reputation and perfectly structured examples of urban narratives based on heritage and local references.

The motivation to pick these three examples, comes from the believe that a sustainable fashion movement can exists within the intersection of modernity and innovation. Any design, material or aspect of fashion items should be globally relevant, meaning possible to wear in a global context. All the brands highlighted here, succeeded t create mindful fashion collections, which are timeless, wearable and still with a strong link to a specific handcraft.

These brands are based in Colombia and India, both regions where the textile craft is very much embedded in the entire culture of the country.

In Colombia we can see that almost one million handwork's support their livelihood directly or indirectly through the worldwide acknowledged and handmade sector. An approximate of about 350,000 of theses people are the artisans which produce in often rural areas of the country their craft in handlooms, knitting and crochet techniques [5]. India has a long standing tradition of the artisan skills just like Colombia has. The handloom and handcraft sector is the second biggest right after the agriculture sector. The Indian artisans are estimated to be almost 23 million today [6]. All of the designers and creatives I talked to had the shared vision of their heritage as one of their main strengths and signature powers. They understood the comprehensive need to build a bridge between the old and the new to use their heritage as an advantage on new international markets.

When looking into this particular bridge-building between traditional textile crafts and new markets, we have to consider the account on which the global aesthetics of fashion is built on. The system of what is perceived modern, beautiful or fashionable is a long-established and mostly westernized construct, catering to colonial mindsets and an artificial force of homogenous appearances. For the longest time brands from all over the world kept resembling a certain kind of composition, validation, or experience that comes from a mostly Eurocentric and homogenous mindset. Imagining that outside of the domestic markets, brands could become a success story without subscribing to the values of these established aesthetics was unthinkable. Mass consumption as well as mass production drove the idea of cookie-cutter designs and looks, widely available for everyone and without room for unique tendencies. Decolonising fashion is a part of the new development in re-learning the value of handmade textiles, it also is one of the now most powerful marketing tools [7].

3.1 BLONI—Education and Innovation to Find Your Customer Group Between Two Worlds

Established in 2017, Bloni we is modern ethical luxury brand which transcends clothing beyond creativity and imbibes it with a purpose that is unique to their clients.

Their creative process encourages ideas and allows their customers to understand and create a self such that the garments become an extension of their personality. With unique handcrafting methods and using traditional techniques practiced via sustainable processes, the brand created a distinguished and unique look.

Bloni launched its flagship store at Chattarpur, New Delhi in November 2018 where it offers its clothes to a wide range of customers.

Akshat Bansal, the founder and creative director has studied at the worldwide prestige Central Saint Martins in London and the well established National Institute of Fashion Technology in India. Akshat trained as an apprentice at Saville Row, London and has honed his skills designing couture at Tarun Tahiliani. He is inspired by unconventional visions and aims to manifest his perspective with Bloni.

This particular time was one of the most influential and mind opening periods in his career. Not only on a skillset level but also in forming the ideas and ideologies which should become so defining for Bloni. When Akshat Bansal studied in London at the prestigious university he also did an apprenticeship, at the iconic Savile Row. He describes that during this time he figured that there is this void between the old and the new. The old way of thinking at Savile Row, where clients would come in and have hour long fittings several times to then buy one piece of clothing with is extremely expensive and based on the idea of keeping this item a lifetime long and reputably wearing it.

As the designer sees it right now, most of his clients and especially men did now want to spend several thousand on one suit only, he deliberately highlights how especially in his local surrounding the idea of repeating an outfit is still met with judgment by the immediate social circles he experiences. Bloni believes in ethical conscious luxury, highlighted by the melting of the craft with technology, making traditions amalgamate with tech future to make relevance and a global experience that is exceptionally unique while continuing to sustain a craft and a design language that handholds the importance of craft. The brand has been constantly working with Econyl—regenerated marine plastic waste. The designer combines the innovative and low-impact materials with the ancient craft of tie-dying to create his own and contemporary relevant design language.

Innovating Local Craft in India (Example: Bloni)

For Bloni, the story of sustainability and handwoven textiles in India can only go together with a bigger strategy and mission to create opportunities for oneself and educate about the impact fashion design can have on matters of sustainability. The designer explains that when he started thinking about sustainability the first stage was concerned about the use of water that every textile production cycle is bond to. As Bloni collaborates with artisan groups for the production of their luxuries garments, the designer is aware that even when working with handmade textiles the brand is still using resources, creating materials, and therefore contributing to the global pollution through the industry. The concern of the designer on worldwide prominent issues of environmental destruction is especially nuanced because of the effects this can have on the surrounding noticeable in India.

Many of the artisans he collaborates with are living in parts of India which might get submerged in water within the next 10 years, this pressuring fact constantly challenges the designer and creative director of the brand to ask himself how to safeguard the regions and contribute to a long-lasting change for his country and the textile industry.

Obtaining the balance and gaining business structures that create a better ecosystem is build into the DNA of Bloni. The concern to not only create a profitable business but also level the chances and have regard for natural resources and living spaces is the most motivational drive of the brand. The inescapable facts of the global climate crisis or harmful ocean pollution are matters which are in the designer's opinion mostly only accessible to a certain level of education. He feels that a lot of the population around him, which for example is less traveled or less exposed to international trends and inputs simply is less aware. He gives the example of his mother, someone who lives by sustainable paradigms in terms of reusing resources, recycling, and being knowledgeable about how to handle fabrics and garments properly. This is a mindset embedded in her culture and acting, but nothing that is ultimately linked to a greater purpose or awareness of the destruction that or planet is facing. The argument he makes is one of understanding that it does not necessarily help to constantly repeat the sustainability message to the same target group and people. 'Telling educated people to get more educated makes no sense.' The designer said. He considers that as fashion designers or as a community it has to be about increasing the various aspects of awareness at the right spot. The key is the empathy for the more rural areas, for people with a less metropolitan mindset and encourage them to keep more slow ways of consumption and resource using alive.

This is not a new concept to teach, it is a going back to values that come from a more respectful and careful society and connection to consumption. The same tactics can be applied to the respect and imagination clients do need to have for textile crafts. Many handwoven textiles back in the days were produced in careful manners until the machinery fabrics emerged and producers and consumers fell into the cycle of faster production of continuously increasing volumes. For Akshat Bansal the education trajectory of the impact of fashion has the same account and direction as the call for respecting handwoven textiles. Introducing transparent communication and conversation around the way he designs and produces is for the designer a way to create a niche and a client group which otherwise would be missing for him and Bloni on the local Indian market.

The Actual Value of Fashion Versus The Social Status (Example: Bloni)

The general analysis of buying fashion and textiles because of their social conjunction and the status or because of their actual worth. Big global fashion players define the value of an item as not necessary anymore by the quality, the definition of what is reliable to invest in is made by status and image. The main developer of the growth and verification of brands and smaller design companies can be driven by marketing tools and even social media influences. These parameters make it incredibly hard for a designer brand like Bloni which is defined by technologically advanced textiles

blended with hand weaved textiles, to persuade clients and business to business partners why their product is worthy. The globalization of brands and the continuously
widening availability of higher price or affordable lucky brands in India is a competition Bloni atelier has to face daily. Being up against international designer labels
and having to justify the price of a locally grown and slowly made item is part of
his entrepreneurship, Akshat Bansal says. The designer reckons that once buyers
see and understand the comparably long process of material innovation and the time
calculated for manufacturing a piece they can be convinced though, and one very
important realization is that if clients do not see the value of his items they are simply
not the brand's target group.

The moving dynamic of the India-based fashion brand is shaped by the vision
and ambition of the founder and creative director. Understanding the push back and
the need India still has as a growing market for sustainability he has an international
outlook to the manifestation of his business, but also the manifestation of ethical and
sustainable textile production in India. First of all, it is clear to Akshat Bansal, that to
develop and have an impact his aesthetics and designs need to be globally relevant.

The route and journey to get into major international markets and shops are most
likely to be shaped by contemporary design and innovations. The audience for his
sometimes completely handwoven textiles and collections has to be able to wear
a shirt that is made with artisan Indian textile skills in a modern-day global urban
context. Only then his brand, his business and his vision to give back to the artisan
communities he is working with will have a real-life ground to stand on. Pushing
the boundaries of design and tradition to be able to function and be decoded on two
different markets as a creative porch many Indian designers are were not willing to
take, it is a recent development of high significance for the domestic Indian market
and the Indian diaspora all over the world.

This process is led by the need to innovate traditional skills in a still respectful
and thoughtful way. The Delhi based designer illustrates how he seeks an active
dialogue and interaction with the artisan groups in his hometown to form business
opportunities and forward-thinking strategies. The gap between what artisan is often
compensated for their labor and for how much fashion items are sold is immense.
In order to close the gap and collaborate on a better level, the designer decided to
bypass the middle man and form his one way of working with handloom experts, the
result will be an upcoming menswear collection made of 90% handloom fabrics.

Generating Impactful Social Change and Material Innovation (Example: Bloni)

This way the designer can navigate through the process of truly generating impactful
social change and material innovation. Simply switching from cotton to silk-based
yard crafted a whole new upscale and elevated look and quality for his pieces. The
option to have more control over order volumes and also train artisan handworks to
deepen their skills in a more contemporary direction is a Win–Win for both parties.
The microsystem he created by working with the weavers directly additionally allows
them to become an active part of the production set up for Bloni. Encouraging them to
get out of their comfort zone and explaining the serval layers of pricing and business
opportunities did help the brand to continue this important and time-intensive work.

For the designer, it is clear that there has to be a larger motivation for growth for the artisan communities. If they will not receive higher order numbers in terms of fabrics, most communities do not have the motivation to adapt to change and learn new skills. If the possibility is given to increase the quantity of the orders by the designer season by season this is a motivation to keep the craft alive and developing for everyone but especially the weavers. The hand loom experts and artisans are actively involved in the process of creating fashion items and shaping their own part of the economy. With forming closer relationships to their clients, they as suppliers can stabilize the accruing order flows and business perspectives.

As we can see this is a context and a cycle which by its nature empowers itself. Innovation means the development of materials and the possibility to grow on international markets but this innovation means also giving the opportunities to keep heritage techniques alive by uplifting abilities and translating them into contemporary structures. The bypassing of the middle man and taking time to develop a solid and meaningful connection to suppliers and artisans also affects the margins that brands and designers do otherwise have to add to their final price calculations. An extra position will affect the concluding price of the item, which in the end could make the difference for the end consumer to take a purchasing decision.

What can be seen in this example is how one single brand can shape and influence the long-overdue set up for a system of exploitative handwork and distant supply chains. If companies cooperate closer with artisan communities the possibilities to move forward with the craft are significantly higher than when orders are just made in a distant business manner. The warning process for both parties, plus the product value for the end consumer increases, and luxury and quality have a chance to be redefined. By forming your won clusters of manufacturing artisans and educating your client group, brands have the china to be globally relevant and still honor their heritage and traditional textile skills.

3.2 A New Cross—Realising that Your Heritage Can Become Your Strength

A NEW CROSS is the story of Nicolás Rivero, the designer, and CEO of the Bogotá based fashion brand. The label was founded with the desire to explore and mx multiple sources and as well as techniques and knowledge to understand and reinterpret traditional methods. The design house is driven to question and deconstruct contemporary ideas on what it means to be Latin-American and how to create a link between contemporary design and true heritage. They have created a minimalist, monochromatic, and urban proposal inspired by the future of Colombian identity while incorporating local heritage through their work supported by several groups of artisanal communities. Craftsmanship techniques that transmit a message of artistic identity that can only be fully appreciated when garments are worn.

Looking into the sourcing and creating processes Nicolas Rivero is working with, and his unique artistic and cultural method gave me the chance to get unparalleled insights into his design choices, motivations, and business strategies. All are balanced between the need to grow and the motivation to perceive the culture of textile skills and artisan knowledge. A New Cross uses about 40% of handwoven fabrics for all of their collections. They are either made in the loom or hand-knitted. The brand has a different division that works exclusively with artisans, divided by some specialized with the difficult handloom techniques or with simpler needled knitting. The artisan groups are located in several parts of Columbia, both very local to the brand and design studio in Bogotá and seldom also in other regions of the Latin American country.

Nicolás Rivero states that Colombia is not practiced in producing or using fine yarns or specific breeds of sheep for their textile and wool production within the country. Places that are as high as the Andes are a perfect ground for certain breeds of sheep to source finer wool but, but still Columbia as no Alpacas or Lamas which you would find in lands like Peru, Bolivia or Chile. The limited availability of more delicate or elaborate fibers forced the brand to also source materials outside of their domestic market. Most of the animals are from other countries, and most of the natural soil sourced fibers are from different regions of Colombia. The largest of the Colombian yarns are made in cotton, a quality that offers the perfect touch and quality properties to the brand's collections.

Handmade Textiles as a Business Decisions (Example: A New Cross)

When asking the brand what is their first motivation for using handmade textiles, the founder immediately mentions how this business decision is based on connection, honesty, and a craft that he respects deeply. A big upside of the process is the level of control a smaller brand like A New Cross can create offer their production ways. Working with smaller artisan groups allows them to get to know every single person that works with them. They are automatically involved in most of the weaving process and see how the garments they create get produced. For the designers, this part is a crucial and very significant part of his creative process. He takes time to go and meet craftsmen, with the artisans and the local people who pick and produce so much of his clothing. Nicolás Rivero sees a connection between the quality and sense you can have for handmade products, which is not translatable to any experience a replicated product in a huge department store can give, especially when it is clear that a product is being mass-produced and is meant to be thrown away after a short amount of time.

For the design and aesthetically approach of A New Cross, the hand weaved and handloom fabrics make a big part of the signature style. The sense of a mistake, the uneven turn that the hand has cannot be built or duplicated by a textile production machine. The Japanese concept of 'Wabi Sabi', the worldview based on the acceptance of the imperfect is what Nicolás Rivero revers to, and what gives him creative imputes as much as freedom and uniqueness. A New Cross aims to tell the story behind their textiles and behind their designs through a piece of fabric and a culturally respectful approach.

For the Bogotá based designer, the crafts of the textile are having a small revival but it's not quite enough and as prominent as it deserves to be in terms of appreciation and global relevance. For him, the new generation is way more aware of what they consume, and what an impact their buying decisions can have on the world around them.

Conscious consumerism and a global view on local issues and retaliations are key to newer and more healthy perspectives. This is a growing trend for the designer and even almost a movement, something that is leading us to be aware of how can we make a change and how we can consume more responsibly. Like every brand, A New Cross is also a business though and besides their creative and moral values, the fashion house aims to have an international impact and a client base that is convinced and loyal to the brand's garments.

Encouraging Consumers to Invest in Long-Lasting Fashion Items (Example: A New Cross)

One of the things that the designer encourages within his clients is that they buy the things that they need and value in their closet. The designer sees that this way little by little he can contribute to a change because he supports his clients to change their mindest from mass production and senseless consumption to a curated and respectful way of buying. The currently still mainstream mindset of neglecting the reflection on what is the human cost of consuming such products, and this also costing to the world is something he wants to move away from.

He refers to the idea of buying things just because they are cheap and not acknowledging who's may be suffering to price it for us to be able to get like a $5 t-shirt as a true reflection of bad textile and business practices and the constant consequence of poor quality products. Like many fashion brands of higher price categories, A New Cross prefers to have loyal clients and clients who understand the worth of the design houses' craft. Some of the band's customers own their designs for ten years, and they still have pieces from that moment when the brand first launched. The durability of the products is a vital and satisfying fact to the design house and a true enabler for sustainability because these pieces have to be respected and maintained by their owners to keep them in a good shape and condition.

The immense regard that consumers need to have for the artisans and for the materials that can be a natural source is still lacking in the bigger picture though. To push that agenda of buying more consciously, it is important to incorporate heritage skills into contemporary design and modern brands. One other very interesting aspect the designer highlights is the take on artisans' handwork and how this needs to be protected from exploitation. The construct of taking the knowledge of a community, and sell it as your own is something to be criticized. Cultural appropriation cannot only happen with a distance to the source of the heritage, it is a power dynamic that often lies within capitalist frameworks and adds benefit to colonial structures. A way to approach working with artisan groups that focus on handwork, weaving, or loom techniques is to openly establish a dialogue with them so it almost becomes a co-creation.

Supporting Local Communities Through Handmade Textiles (Example: A New Cross)

According to the designer, Colombia is a country with many political issues, a certain level and stigma of poverty. That poverty has made its way to artisan groups and communities around the Latin American country. Their struggles and basic needs to make sure they, their families, and their communities can survive are to be considered when establishing any kind of work relations.

A first built relationship that gives the brand the chance to observe local techniques was the starting point for a longer collaboration between A New Cross and several artisan communities. From there the funder Nicolás Rivero was able to begin to learn very respectfully, how the handloom experts administer their craft. Only this introduces the baseline to then narrate with the process and potentially begin to start implementing distinct new ideas on how the artisans can use their knowledge and add different techniques or materials to it to create new, modern and unexpected twists of fabric and materials. This can be a learning process and a very opportunity-rich way of working for both, designers and brands but also handloom and knitting specialists.

For A New Cross, this has been a revelation to become linked to their local artisan groups and Colombian heritage.

New Aesthetics and Heritage s Kills as a Unique Selling Point (Example: A New Cross)

The founder Nicolás Rivero described how special the brand is in terms of aesthetics and how his designs first got noticed on the global market and at fashion destinations as Paris Fashion Week. When you see a product of A New Cross, you can feel all of the ethnicity and the hand labor behind it, but it is not a souvenir. It is a design piece, representing the melting of two visions and narratives. There is the very contemporary minimalistic functional design choice and this combines with all of the crafts and all of the knowledge, and all of the material that speaks in a language that's very honest and heritage driven. Especially when the brand began to sell on the European market, the responses from clients, buyers, and consumers started to evermore be of surprise. Nearly every A new Cross garment says 'Handcrafted in Colombia,' on the label and for the Bogotá based designer, it is of high importance to show where the pieces were made.

Rivero believes, that the reputation that Colombia as part of the global south has, is often chapped by medial anecdotes or cliches. When we think of Colombia as a country the first impressions we consider are the country's crime rate, soccer, coffee, and something very tropical very floral. All of the brand's designs are mostly black or very monochromatic, a representation of the other face of Colombia. This way the founder follows his mission to essentially educate consumers and the world in how Colombia is so much more than that.

Colombia also has beautiful cities and modern urban structures. With an enriched cultural offer, something that lives and keeps arts, gastronomy, and people inhabited. The brand tries to talk about all of those things through the textile because A New Cross believes fabrics or garments are the smallest space you can inhabit.

These garments or these architectures not only keep you from being but they also communicate the stories that need to be told to show a diversity of textiles and fashion.

Nicolás Rivero found with his brand the perfect combination of heritage in a contemporary way, which represents the old and the new, and instead of trying to fit into an already prepared path, defining your brand's strength in the roots of its origin.

3.3 Priah—Social and Entrepreneurial Growth Through Handmade Textiles

Priah is a Columbia based fashion brand specialized in the intersection of heritage and luxury items. Their mostly hand-woven or crochet garments can Juanita García the founder and head of creative for the fashion brand has a long-lasting and strong relation to handmade textiles and the craft that is deeply anchored in the Colombian country around her. The brand is veneer collections in the limitations of the items and how the garments can function as a canvas to highlight the value of hand crochet and handmade textiles. The brand has a mission to teach this tradition to new generations, and disrupt the lack of demand at fair prices for handmade fabrics, to complete the cycle and make it desirable for younger folks to keep sourcing locally.

Their motivation and inspiration comes from the goal to give handmade textiles and motivation for using handmade crochet textiles are to give traditional art space in a modern industrialized world. To create empowerment and endorse the hope of women artisans who want to be acknowledged and respected for their talent. And especially to give a voice to so many hands which craft has been taken for granted. The brand does not use the same amount of artisan-made and crochet textiles in every garment.

Due to the time it takes to hand-crochet any of their designs and textiles with extra-fine yarn and ta specific crochet technique, it would be impossible to a full collection just made by artisan communities. Smaller pieces can take up to 07–25 days on average and sew them together with special natural fabrics in contemporary garments.

Colombian Crochet and Minimalistic Designs (Example: Priah)

As Juanita García says 'the result of doing it back up our idea of finding a balance between heritage inspiration and minimal aesthetics.' The crochet fabrics locally produce and locally sourced; handmade by grandmothers and mothers artisans from Boyacá in Colombia. Although the technique we use, named Granny Square, is not a local invention; the diagrams we crochet are inspired by Muiscas' indigenous folklore art. The designer told explained that 'when we designed our first collection, the fact of replicate and co-create based in their sacred geometry made sense for us. We wanted to acknowledge the last time we had an authentic identity.'

Dynamics and Challenges When Working with Artisan Groups (Example: Priah)

To understand the way Priah collaborates with artisans, I looked into the structure and requirements the owner sets to establish and determine the groups she preferably does work with. Juanita García explains how at the beginning each of her projects mostly started by finding groups and charities. These charities of artisans usually perform any kind of textile handwork to support themselves and keep the profession alive. She started and hired a group of 17 women all of them in vulnerable positions as being elderly people with very low income. To Juanita García it was a logical step and decision to contribute to their lives by offering them employment. On the forefront and for any kind of marketing approach this is undoubtedly a perfect and excellent story, but behind the scenes, it was difficult to work with the chosen artisan group.

The impression the brand's founder had was that because of a certain dynamic that can get established when working with indigenous groups, it can be difficult to navigate through business interests. Sometimes it could have happened that specific work deadlines would be pushed back for weeks, calls were left answered, and provided raw material was left unused or wasted. For Juanita García this was an alarming signal, that to keep operating with artisan groups and keeping the relationship to her heritage and the local textile craft, she as a brand and employer had to do more than just hand in production orders. Initially, this started with a small group of women who showed a collaborative mindset and were aiming to support themselves more stable and ongoing through their work.

She chooses the personalities who had the values she wanted to work with and encouraged them to start communicating with people that have similar visions for their work and life, to build momentum and recruit more and more women. The owner of the brand has a strong educational background in political scientist, finance, and social entrepreneurship, which helped her to build this brand as a bigger mission and impactful narrative for her region. She saw the necessity and direction this project needed to make a social impact and move into better fashion futures.

Once the group started growing Priah began to roll out training for their hand workers. They were not only encouraged to deepen their skills in crochet and handwork but also to learn entrepreneurial skills. The founder Juanita García encouraged the artisan community she founded to sharpen their business senses and see themselves as artists. It is very difficult and special to find 'hands like theirs' and therefore they have to understand their value and meaning as craftsmen, she said. An honest and transparent approach to commercial value and sales strategies was also part of this learning process.

Fostering Entrepreneurial Knowledge Within Artisan Groups (Example: Priah)

Priah explained and showed examples of what does happen with handmade fabrics. How the crochets or weaves get implemented into their contemporary garments and designs and afterward sold in upscale retails stores or online shops. The equation of how a garment that sells for 500 Euros, is calculated by the production costs, ecological materials that might be mixed in, plus marketing and creative direction budgets trainees to establish the value of their profession. The women learned and experienced a whole new outlook on business calculations and how they can sell their products for the right prices also to other clients.

For the Colombia-based brand, this is a heartfelt project which keeps growing and enveloping over time. As the founder mentioned, the Colombian government through Artesanias de Colombia has done strong efforts to support the local artisan heritage. However, she is critical if the basic idea of subsidies, and pieces of training to just primarily teach the handmade skills will help to stabilize the social and economical struggles artisan communities are facing. The opportunity to incorporate the craft into new designs and aesthetics to continue an old tradition seems to be a far more long-lasting and efficient way to go forward.

The process took around five years until Juanita García felt that Priah had a team of manufacturing in which the brand had the right level of quality and strategy to achieve a healthy business construction for themselves and the women that they support. When asking the founder if she can see the impact of her educational program and the work she is offering, she defiantly reveals several layers of positive change. One very simple fact is the social dynamic and giving women the chance to connect with each other and meet for a work session, that then automatically also function as a chance to engage and exchange with each other.

The Social Dynamics or Supporting Craftsmanship (Example: Priah)

Especially, considering the fact that a lot of these women are mostly stay at home mothers or wives, this is an experience that enriches their lives and gives them purpose. The designer recalls how one woman, in particular, expressed the change and impact her new set up has for her whole family, she is able to contribute to the household's economy, a huge step in feeling independent and worthy. Juanita García observes that in Colombia, or Latin American cultures are many layers of social and financial class, and most countries have growing gaps between the income of the population. The chances you become to grow your skillset and better a situation are very much linked to the privilege of education.

Considering the aspect of the client base and how fashion is distributed and valued on a global market, Priah is building the bridges between people and objects, motivated by the need to solve a problem. The designer seems a chance in the consumer's mindsets, how now products aren't purely bought to serve a function but also to represent values of the person wearing the clothes. Artisans know how to create individual and valuable items but can often lack the skills to translate this into a modern concept.

For the brand pairing artisanship along with design for a contemporary world can make the difference between making a living out of the craft or not. It can help the artisans to access niche markets and clients who more likely to pay higher prices for crafted items and hand-made garments and it can expand their access to explore different materials. All of these aspects will ultimately support artisans to differentiate from competitors and sharpen their curiosity for and elevated authenticity. By using these practices the founder and designer Juanita García wants to make sure that there has to be a level of carefulness when collaborating with artisan groups.

Especially the indigenous community is one of the least in favour of these parameters, a systemic and culturally anchored disadvantage that needs to be worked against and balanced out. While for the brand it is important to create spaces of social equality and connection the founder wants to point out that simply 'helping' is not enough. Structures and the distribution of knowledge have to be introduced in order to fully empower people and create respectful business opportunities for everybody.

4 Conclusion

What the last examples showed us is that implementing textile heritage skills, artisan communities, and a general slower combination of hand manufactured textile and other sources is a valuable practice for modern-day brands. It supports textile artists and groups by keeping the craft alive and surely is a form of enriching the textile industry. Endorsing artisan handwork is an approach that needs to be introduced together with a connection to make sure that no simple plagiarism or exploitation is supported or executed.

Creatives who are committed to this project and understand how to operate this business in a flexible and resourceful manner will shy away from just copy and pasting designs and aesthetics. Their businesses are hardly ever built on mass production and have instead key signatures to them that are often rooted in the individual experiences of the designers and founders. There is a real chance in sustainability and sustainability culture for the textile and fashion business is to learn from these experiences and skillsets. An enhanced connected journey has to be at the forefront of this movement towards sustainability. This cannot only be done by one player, this cannot only be done by one part of the supply chain. It has to come with association and connectivity.

The three brands which have been the main focus of this case study have been highlighting the described way of working and establishing a supply chain with all of their positions all of their ways to act shaped by their own immediate experience and expertise. These viewpoints are surely defined by parameters as their customer groups, the established character of the garments, design and visual choices and preferences, and the own values and motivations of each designer and their teams. While these are potentially not the perfect constructed solutions for everyone and every part of the supply chain, it is an inspiration to see forward and keep forming kinships between artisans and contemporary designers.

This has to be seen as a pledge for a healthy fashion future and for a non-exploitative way to work with artisan communities. The knowledge that we can combine with abilities helps to build foundations for sustainable storytelling, a part which every company or brand is relying on to grow. When we learn from artisan collaborators and respect cultural dynamics, we can implement this to fashion branding for a global audience. This can in the end help to educate consumers and create an awareness on the global fashion market. Learning from indigenous manufacturing processes and views on durability is the essence of slow fashion. Contemporary slow and handmade fashion elements will become a chance to support brands, regions and a healthier environment.

References

1. Donaldson T (2020) Is slow fashion the new luxury? WWD. https://wwd.com/fashion-news/ready-to-wear/slow-fashion-artisan-luxury-sustainability-social-impact-1234575062/.
2. The State of the Handworker Economy (2018) www.buildanest.org, The Nest, p 26. https://www.buildanest.org/shereport/.
3. Backs S, Jahnke H, Lüpke L et al (2020) Traditional versus fast fashion supply chains in the apparel industry: an agent-based simulation approach. Ann Oper Res. https://doi.org/10.1007%2Fs10479-020-03703-8
4. Granskog A, Lee L, Magnus K-H, Sawers C (2020) Survey: Consumer sentiment on sustainability in fashion. https://www.mckinsey.com. McKinsey & Company. https://www.mckinsey.com/industries/retail/our-insights/survey-consumer-sentiment-on-sustainability-in-fashion
5. Wipo.int. (2019) Arts and crafts of Colombia. https://www.wipo.int/wipo_magazine/en/2006/06/article_0002.html
6. Banik S (2017) A study on financial analysis of rural artisans in India: issues and challenges. [online] Banik S, A study on financial analysis of rural artisans in India: issues and challenges (November 15, 2017). Int J Creative Res Thoughts (IJCRT). https://papers.ssrn.com/sol3/papers.cfm?abstract_id=3137936.
7. Jansen A (2019) Decolonising fashion | Vestoj. Vestoj.com. http://vestoj.com/decolonialising-fashion/.

Consumers' Attitudes Toward Sustainable Luxury Products: The Role of Perceived Uniqueness and Conspicuous Consumption Orientation

Andrea Sestino, Cesare Amatulli, and Matteo De Angelis

Abstract This chapter investigates the effectiveness of luxury brands' messages focused on product sustainability rather than on traditional luxury product features. The study sheds light on the role of perceived product uniqueness and consumers' conspicuous consumption orientation. An online experiment has been conducted among a sample of 144 participants exposed to a luxury product communication message focused on product sustainability or focused on product performance. Results show that a communication message related to a luxury product and focused on sustainability (vs. performance) leads to higher positive attitudes toward such product and this effect is mediated by consumers' perceived product uniqueness. Additionally, findings underline the role of conspicuous consumption in magnifying such effect. Implications and suggestions for marketers and managers are discussed.

Keywords Sustainable development · Sustainable luxury · Sustainable consumption · Perceived uniqueness · Conspicuous consumption · Luxury brands · Luxury marketing · Accessories

1 Introduction

Sustainability has been globally recognized as a central issue by companies and consumers [1]. Indeed, companies are gearing up to make significant changes to their activities, accordingly with the emerging challenge about balancing sustainability with profitability and brand appeal [2]. This is particularly truthful in the

A. Sestino · C. Amatulli (✉)
Ionian Department of Law, Economics, Environment, University of Bari, Bari, Italy
e-mail: cesare.amatulli@uniba.it

A. Sestino
e-mail: andrea.sestino@uniba.it

M. De Angelis
Department of Business and Management, LUISS University, Rome, Italy
e-mail: mdeangelis@luiss.it

© The Author(s), under exclusive license to Springer Nature Singapore Pte Ltd. 2021   267
M. Á Gardetti and S. S. Muthu (eds.), *Handloom Sustainability and Culture*,
Sustainable Textiles: Production, Processing, Manufacturing & Chemistry,
https://doi.org/10.1007/978-981-16-5967-6_12

context of luxury, where managers must be able to keep the emphasis on prestige and craftsmanship by combining them with sustainable values [3], in the attempt to make such product unique, and recognizable. Despite the relevance of sustainability for luxury brands and consumers, past research has mainly focused on sustainable consumption in the context of mass-market products and brands [4]. Such new challenges and opportunities could not be rest unconsidered, especially by considering how the luxury market is steadily growing and showing new opportunities for companies [5], coherently with new emerging consumers' needs and desires [6], and with the new meanings that luxury consumption is assuming (Seo and Buchanan–Oliver, 2019). Indeed, the total amount of revenue in the Luxury Goods market amounts to US$ 285,137 million in 2029, and the market is expected to grow annually by 6.4% accordingly a CAGR 2020–2025 (Luxury Goods Report, 2020). This revenue growth also coincided with a mounting commitment toward sustainable development among luxury brands as for Gucci and Porsche, who launched some eco-friendly products (e.g., Gucci's zero-deforestation handbags made with Rainforest Alliance-certified leather and the Porsche Cayenne S E-Hybrid, respectively). In fact, recently, the major high fashion houses have realized that they must be the first to open the way to sustainable fashion: this is the case of Vivienne Westwood who in 2019 launched the "Buy Less, Dress Up" campaign based on the concept of cruelty-free, offering an avant-garde collection free from fur and animal skins (https://www.vogue.co.uk/gallery/vivienne-westwood-spring-summer-2019). Again, Stella McCartney, together with Adidas, launched the new Stan Smith vegan sneakers by Stella McCartney, therefore without animal leather (https://www.stellamccartney.com/experience/it/celebrate-the-movement-with-the-new-stella-stansmith/). Dior instead proposed the spring–summer 2020 fashion show in a wood, created ad hoc thanks to 164 trees as part of the #PlantingForTheFuture initiative, whose shrubs that were then replanted in specific areas of Paris, symbolizing the need to always be more united in the fight against climate change (https://www.vogue.co.uk/fashion/article/suzy-menkes-christian-dior-catherine-dior-maria-grazia-chiuri).

However, despite the wide recognition of sustainability by the major luxury companies, and the business relevance of sustainability for luxury consumers, studies only investigated sustainable consumption in the context of mass-market products and brands (see [4] for review), with a return of a lack of data about what underlies consumer reactions to luxury goods that are somehow characterized by sustainability elements. Generally, consumers' attention focus on a negative and positive perspective: The negative perspective describes luxury and sustainability as two incompatible and sometimes conflicting concepts, to the point that sustainability may undermine consumers' attitudes toward luxury products [2, 7, 8]. The positive ones suggest that consumers might sometimes favorably assess luxury goods that are somehow sustainable, e.g., [9–12]. Moreover, in this field of studies, there seems to be a lack of contributions about consumer reactions to luxury goods that are communicated with a focus on their sustainable characteristics [13]. Currently, consumers are becoming not only interested to the performance of the product and its intrinsic features but also sensitive to products' attributes in terms of environmental sustainability [14].

The present chapter is aimed to fill this gap by assessing the effectiveness of luxury brands' messages focused on product sustainability rather than on traditional luxury goods features, by shedding light on the role of consumers' perceived uniqueness of such luxury products. More specifically, we conducted an experiment that investigates consumers' reactions to luxury products communication messages that are sustainability-focused versus messages that are performance-focused, by revealing how communication message related to a luxury product and focused on sustainability (vs. performance) will lead to higher attitudes towards such product and this effect is mediated by consumers' perceived product uniqueness. Additionally, the chapter also investigates the role of an important consumer-related characteristic that is conspicuous consumption orientation: we highlight how the effect on consumers' attitudes is also influenced by consumers' conspicuous consumption, as the ostentatious display of wealth to indicate status. In particular, the study sheds light on the counterintuitive underlying mechanism that may lead consumers to have more positive attitudes toward a luxury product communicated through an emphasis on its sustainability than a luxury product communicated through an emphasis on its performance-related attributes. Results show that communicating a luxury product by focusing on its sustainability (vs. performance) may activate a higher level of perceived uniqueness of that product, which in turn leads to positive attitudes toward it. Such results are in line with previous studies suggesting that uniqueness is one of the most important drivers of luxury purchasing [15, 16]. Interestingly, results underline that the effect of sustainability-focused (vs. performance-focused) communication messages of luxury products is moderated by the level of conspicuous consumption orientation which characterizes consumers. In particular, the higher the level of conspicuous consumption the weaker the effectiveness of sustainability-related (vs. performance-related) messages. We discuss the theoretical foundation of such effects and test our hypotheses through a moderated mediation analysis. We contribute to marketing academic literature in three different ways, by showing that emphasizing product sustainable features, instead of product performance, may play a key role in increasing consumers' attitudes toward the product. Moreover, the chapter enriches our understanding about the role of uniqueness in luxury, by underlying that such an important attribute may also be leveraged through a focus on sustainability. Furthermore, the current research contributes to literature on conspicuous consumption orientation by shedding light on the role that this consumer-related characteristic may have in sustainable luxury development. From a managerial perspective, our results may be useful both to managers and policymakers: Indeed, findings suggest that luxury brands may generate more positive consumers' reactions in terms of attitudes, by highlighting product sustainability rather than product performance, in their communication activities. Results also shed light on how luxury marketers and managers could partially revisit the traditional idea of uniqueness in luxury by managing it as a result of environmental sustainability.

2   Theoretical Background

Sustainability is a buzzword today [17], also accordingly to the new meanings that luxury is assuming [18] and its widely recognized relevance for luxury fashion [2]. The Organization for Economic Co-operation and Development (2000) defined sustainability as the consumption that supports the ability of current and future generations to meet their material and other needs, without causing irreversible damage to the environment or loss of function in natural systems. Accordingly, sustainability assumes different meaning and facets, in several field and industries, linked by a common trait: In the field of fashion and luxury, choosing sustainable strategies has been recognized as one on the best possible way to safeguard the environment, workers' rights, and consumer choices [19]. Moreover, greater attention has been direction on products' strategies and related characteristics, according to the concept of sustainability, represented by a new approach to the design of clothes for materials, production processes, and marketing [19]. Indeed, in this sense, the clear role of sustainability emerges, based mainly on the use of non-harmful materials or materials with minimal impact on the environment, both in the production and disposal stages, such as silk or hemp [20]. Interestingly, consumers are becoming increasingly concerned toward sustainable consumption: According to Nielsen's research report (2018), about the 48% of US consumers say they would change their consumption habits to further limit the impact of their purchases on the environment. Thus, given consumers' increasing concern for the environmental and societal impact of their consumptions, most companies while not traditionally devoted to sustainability, have come to realize they can no longer ignore the topic, including those ones part of the luxury world (e.g., [12]).

Aligned with renewed stimulus for companies and consumers' growing interest in sustainability issues, academic research has been engaging in a very lively debate on sustainable luxury branding and consumption, noting interestingly, how luxury and sustainability are convergent because both focus on rarity [21]. Similarly, [9] highlighted how "luxury is about high-quality products that are objectively rare because they employ rare materials and unique craftsmanship skills. On the other hand, sustainable development is about preserving natural resources by limiting the excessive use of materials that can exceed the world's recycling capabilities" (p. 278), conceptualizing the idea that luxury brands are inherently more sustainable than their mass-market counterparts due to some fundamental characteristics of luxury goods, such as limited production, craftsmanship, and durability. Furthermore, by relying on craftsmanship, luxury companies help preserve traditional jobs that are traditionally performed by artisans located in small villages, thus contributing to the well-being of a territory and its surrounding community. Despite the conceptual positive compatibility between luxury and sustainability, research has suggested that consumers see luxury and sustainability as conflicting concepts (e.g., [2, 7, 8, 22, 23], by demonstrating that luxury consumers tend to regard sustainability elements as factors of secondary importance in their consumptions.

Not surprisingly, sustainability features, when included in luxury products, seem to undermine luxury consumers' perceptions of goods' overall quality [7]. Similarly, [10] demonstrated that luxury consumers might prefer to buy an unsustainable luxury good over a sustainable one if they believe the former grants higher status and prestige. Luxury consumers usually look for a sense of great uniqueness in their consumptions [24–26]: Such consumers' need for uniqueness is defined as their tendency to pursue differentness relative to other consumers, through the acquisition, utilization, and disposition of consumer goods to develop and enhance one's self-image and social image [27]. This is also coherent with the paradigm of internalized and externalized luxury consumption, in which consumers usually consume luxury both to signal their status to others (externalized) and to satisfy their internal needs (internalized) (as in [28]). This concept is also linked with consumers' desire for unique products (Desire for Unique Consumer Products, hereafter "DUCP"), and the related perceived product uniqueness as a construct that measures the extent to which consumers hold as a personal goal the acquisition and possession of consumer goods, services, and experiences that few others possess [29].

Indeed, some antecedents of DUCP include individual differences in the need for uniqueness (Snyder and Fromkin 2002), need for status, ostentations, and materialism. Thus, by extension, we assume that consumers led by feelings of uniqueness could be positively attracted by those sustainability products (as for luxury products), including unique characteristics as handmade products. In the domain of product design [30] highlighted that the use of symbolic (vs. functional) product design enhances consumers' perceptions about the brand's symbolic nature, especially when the symbolic design is incongruent with the product category. Building on this idea, we argue that luxury brands' focus on sustainability might be perceived by consumers as a uniqueness strategic approach. By adopting a sustainability-focused strategy, luxury brands might be seen as deviating from the traditional focus on performance, quality, prestige, excellence, inaccessibility, and aspiration (e.g., [15, 27, 31]. Past behavioral research has defined such improbable events as those that are perceived to be unlikely to happen [32, 33], but importantly, they can also be perceived as extraordinary and unique [34]. Notably, the marketing literature has investigated consumers' preference for uniqueness, showing that they feel greater happiness when engaging in uncommon rather than common experiences [35].

Moreover, as mentioned above, we suggest that uniqueness could be shown by consumers to others, by leveraging luxury purchased products, according to the theory of conspicuous consumption, which ties luxury good consumption, with the mere function of ostentatious display of wealth to indicate status, as originally defined by [36]. Specifically, literature showed how individuals led by luxury consumption, maybe proneness to show others their status, and more conspicuous-style consumption (Eastman and Eastman, 2005). Such concepts referred to consumers' status consumption orientation, and conspicuous consumption sometimes seems to overlap, because linked by the common trait of showing others their status, by leveraging on luxury products purchased, and perceived as identical [37, 38]. To clarify, we acknowledge the concepts that they are not exactly the same because such constructs

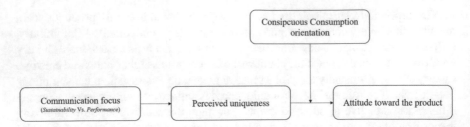

Fig. 1 The proposed conceptual framework: The role of perceived uniqueness and conspicuous consumption orientation on consumers' attitude toward the product

capture different shades of a similar phenomenon [39]. Accordingly, [40] demonstrated that status consumption and conspicuous consumption are distinct constructs that reflect a different set of consumer behaviors and consumption motives: Particularly, while status consumption is mainly referred to consumers' behavioral tendency to value status and acquire and consume products that provide status to the individual, while conspicuous consumption is mainly referred to consumers' tendency to enhance their image, through overt consumption of possessions, and communicating such status to others. Thus, consumers' proneness to show something, differently from making status as recognized, can be captured by the idea of conspicuous consumption, mainly focused on the product visibility [41–42]: Accordingly, such consumers' conspicuous refers to the tendency to buy symbolic and visible products to communicate distinctive self-image to others [43].

Based on the above, we suggest that investigating consumers' responses after exposing them to communication messages portraying luxury brands as focused either on environmental sustainability or product performance could be influenced by the perceived uniqueness of the luxury brand in developing such products, also by considering consumers' conspicuous consumption as their propensity to spending money and acquiring luxury goods, to publicly show their economic power to attain or maintain given social status [40]. Formally, we hypothesize that:

H1: *A communication message related to a luxury product and focused on sustainability (vs. performance) will lead to higher attitudes towards such product and this effect is mediated by the perceived product uniqueness. Moreover, this effect is also magnified by consumers characterized by low conspicuous consumption.*

To clarify, our conceptual framework is presented in Fig. 1.

3 Methodology

In order to provide empirical support to our hypothesis, 144 Italian participants (79 females and 65 males; $M_{age} = 32.35$, $SD = 10.16$) recruited online through the link of a digital questionnaire (see Appendix A) shared on social networks (both Facebook and Linkedin). The questionnaire was divided into three main sections. In the first one, we measured consumers' conspicuous consumption orientation through a 12-item scale drawn by Chaudhuri, (2011; e.g., "It says something to people around me when I buy a high priced brand"; "I buy some products because I want to show others that I am wealthy"; "I would be a member in a businessmen's posh club") on a seven-point Likert scale (1 = "Strongly disagree"; 7 = "Strongly agree"). In the second section, participants have been automatically assigned to one of two conditions within a two-cell experiment that manipulated the focus of a written communication message (sustainability-focused versus performance-focused) related to a luxury unbranded and unisex wallet. Then, participants were asked to indicate their perception of the product perceived uniqueness (i.e., "How unique is the wallet?"; 1 = "Not at all"; 7 = "Very much"), and their attitude toward the product, on a seven-point scale ("On the basis of its characteristics, how do you perceive the wallet?"; "Negatively"/"Positively"; "Favorably/Unfavorably", "Bad/Excellent", "Agreeable/Disagreeable"). Finally, participants answered some demographic questions. Among the participants, the majority (52%) declared to have a B.Sc. or M.Sc. education level, 25% of them a Ph.D. and the rest (23%) of them declared to have low or equal to high school degree (specifically, 22% high school or equivalent, and 1% less than high school diploma).

4 Results

In order to testing our hypothesis, we conducted a moderated moderation analysis. We first mean–cantered the measures of perceived uniqueness, conspicuous consumption orientation, and attitudes. Thus, we employed the PROCESS SPSS macro (Model 14) developed by [41] to conducting our moderated mediation analysis. Specifically, the analysis has been conducted by setting the type of communication (sustainability vs. performance-focused) as the independent variable, the perceived uniqueness as a mediator, consumers' attitude as the dependent variable and conspicuous consumption orientation as the moderator.

The first step of this analysis showed that the sustainability-focused communication (coded as 1) led to higher perceived uniqueness of the luxury wallet than the performance-focused communication (coded as 0; $M_{sustainability} = 32.82$, $SD = 10.57$, versus $M_{excellence} = 31.38$, $SD = 9.29$; $b = 1.26$, $t = 4.77$, $p < 0.001$). Then, we regressed participants' attitudes toward the product on the binary independent variable, perceived uniqueness and consumers' conspicuous consumption. We found that all of them, the independent variable ($b = 0.56$, $t = 2.17$, < 0.05), perceived uniqueness ($b = 0.60$, $t = 2.92$, $p < 0.001$) and consumers' conspicuous consumption ($b = 0.55$, $t = 2.35$, $p < 0.05$) had a significant effect on the participants' attitudes toward the product. More importantly, the effect of the interaction between uniqueness and conspicuous consumption resulted significant ($b = -0.13$, $t = -2.15$, $p < 0.05$). When looking more closely into this interaction, through the analysis of the conditional indirect effects we found that this outcome was stronger for participants with a low conspicuous consumption level ($b = 0.34$, 95% C.I. = 0.18, 0.55) compared to those with a medium ($b = 0.17$, 95% C.I. $= 0.02, 0.32$) and high ($b = 0.00$, 95% C.I. $= -0.22, 0.23$) conspicuous consumption level. Overall, these results lend full support to our hypothesis, demonstrating that the effect on participants' attitudes toward the product of the perceived uniqueness associated with a communication message focused on sustainability is stronger for consumers characterized by a lower conspicuous consumption.

5 General Discussion and Conclusions

In this chapter we explored consumers' perception about sustainable products by assessing the effectiveness of luxury brands' messages focused on product sustainability rather than on traditional luxury goods features: We conducted an experiment investigating consumers' reactions to luxury products communication messages that are sustainability-focused versus messages that are performance-focused. Indeed, while sustainability has been a long-debated issue [44], sustainable luxury consumption only recently captured the attention of academic research [9, 45]. Athwal et al. [10–12], characterized by a perspective in which luxury and sustainability have generally been seen as contrasting concepts [2, 7, 23]. In an attempt to bring these two seemingly distant worlds closer together, we leverage on the concept of perceived uniqueness, showing how consumers manifest higher attitudes toward luxury products after being exposed to communication messages saying that the related brand is engaged in sustainability rather than product performance, in which the sustainable approach in design or productions (i.e., materials, craft), increase products' uniqueness. Specifically, results show how communication messages related to a luxury product and focused on sustainability versus performance will lead consumers to higher attitudes toward such product and this effect is mediated by the perceived product uniqueness: However, by considering consumers' conspicuous consumption, this effect is magnified by consumers characterized by low level of conspicuous consumption.

Thus, our chapter contributes to the literature on sustainable luxury consumption by suggesting that the potential incompatibility between luxury and sustainability might actually have beneficial effects in terms of consumers' attitudes toward such luxury products, when marketers and managers leverage on products characteristics derived by sustainable production process (i.e., handmade) or materials used (i.e. eco-leather), communicating them as unique. Sustainable characteristics in luxury products could confers a sense of uniqueness to consumers, which is widely known to be a powerful driver of luxury purchasing decisions (e.g., [31]). Moreover, our contribution also offers interesting suggestions to luxury managers and policymakers, highlighting how the brand's focus on sustainability in communication messages may earn more positive reactions from consumers than solely emphasizing product performance. To clarify, companies should not neglect product performance but should recognize the opportunity to illuminate their pursuit of excellence alongside their positive impact on the environment and society at large. Furthermore, our results may encourage luxury managers to partially revisit their traditional idea of uniqueness, which can be based on product sustainability and not only on product inaccessibility, prestigious, and rarity. Additionally, we contribute to the literature on luxury marketing by showing that perceived uniqueness converging consumer traits may play a central role in such sustainability versus performance framing in proposing specific communications, considered in conjunction with consumer level of conspicuous consumption. From a managerial perspective, our results suggest, before investing such marketing strategies, luxury managers should seek to understand to what extent their consumers look for uniqueness, and what extent they look for conspicuous consumption. Additionally, luxury marketers should be able to preliminarily segment their markets on the basis of consumers' seeking of uniqueness, and on the basis of conspicuous consumption orientation.

According to our insights, future research could investigate if our findings hold even when comparing luxury brands' communication focus on sustainability with a focus on other common elements. Moreover, we acknowledge that although our results importantly reveal the benefits of highlighting luxury brands' investment in sustainability initiatives, they do not capture whether this benefit changes based on luxury consumers' characteristics as, for instance, in [9] where consumers' approach to luxury consumptions has been investigated accordingly to different motivations (internalized and externalized consumption).

Luxury and sustainability thus seem to become a possible combination. Further to what was suggested, marketers and managers could thus adopt a more open mindset to sustainability aimed at reducing the number of annual collections and perhaps creating timeless collections, using alternative materials without compromising production cycles, and considering the reuse of old materials obtained through fabric recycling programs, upset unique luxury productions (such as Vivienne Westwood's launch of the cruelty-free fashion line, free from fur and animal skins in the 2019 Collation), and contribute to applying rigorous and valid regulations for any operator in the fashion field so as to guarantee a common adhesion to the concept of sustainability. Finally, embracing sustainability does not only translate into an ethical commitment or an economic benefit with important repercussions in terms of investment or substantial energy savings, but it can prove to be an important strategy for increasing the reputation of the company in consumers' eyes.

Appendix A—The Questionnaire (Translated from Italian to English)

Conspicuous consumption orientation

- It says something to people around me when I buy a high-priced brand
- I buy some products because I want to show others that I am wealthy
- I would be a member in a businessmen's posh club

- Given a chance, I would hang a Hussain painting in drawing my room
- I would buy an interesting and uncommon version of a product otherwise available with a plain design, to show others that I have an original taste
- Others wish they could match my eyes for beauty and taste
- By choosing a product having an exotic look and design, I show my friends that I am different
- I choose products or brands to create my own style that everybody admires
- I always buy top-of-the-line products
- I often try to find a more interesting version of the run-of-the-mill products because I want to show others that I enjoy being original
- I show to others that I am sophisticated
- I feel by having a piece of a rare antique I can get respect from others

Manipulations of the communication focus

Experimental condition 1: Sustainability-focused message

This is certainly a luxury wallet, in the finest leather, for both men and women. It is the perfect accessory to hold credit cards and banknotes with elegance. It is an accessory entirely handmade, prestigious, with a classic design.

It is a wallet that has a peculiar characteristic; it is very sustainable. The creation of this purse is based on a manufacturing method that guarantees zero carbon dioxide emissions. As for the raw material, vegetable leather with a very low ecological impact has been developed to guarantee a better and cleaner environment.

Experimental condition 2: Performance-focused message

This is certainly a luxury wallet, in the finest leather, for both men and women. It is the perfect accessory to hold credit cards and banknotes with elegance. It is an accessory an entirely handmade, prestigious, with a classic design.

It is a wallet that has a peculiar characteristic. It is very performing. It is very light, it is made of waterproof material, and it is very flexible. In addition, it is equipped with a space for credit cards, an additional space to store any small paper documents, an internal zipped pocket, and a flat pocket for business cards.

Perceived uniqueness of the product

– How unique is the wallet?

Attitude toward the product

– Negatively/Positively
– Favorably/Unfavorably
– Bad/Excellent
– Agreeable/Disagreeable

Sociodemographics

a. *Gender*

  – Male
  – Female

b. *Year of birth*
c. *Education level*

  – Lower than High School
  – High School
  – B.Sc. or M.Sc.
  – Ph.D.

References

1. Snyder CR, Fromkin HL (2012) Uniqueness: the human pursuit of difference. Springer Science & Business Media, Berlin
2. Kapferer JN, Michaut-Denizeau A (2017) Is luxury compatible with sustainability? Luxury consumers' viewpoint. In: Advances in luxury brand management. Palgrave Macmillan, Cham, pp 123–156
3. Joy A, Sherry JF Jr, Venkatesh A, Wang J, Chan R (2012) Fast fashion, sustainability, and the ethical appeal of luxury brands. Fash Theory 16(3):273–295
4. Lunde MB (2018) Sustainability in marketing: a systematic review unifying 20 years of theoretical and substantive contributions (1997–2016). AMS Rev 8(3–4):85–110

5. Chan ES, Okumus F, Chan W (2017) The applications of environmental technologies in hotels. J Hosp Market Manag 26(1):23–47
6. Oe H, Sunpakit P, Yamaoka Y, Liang Y (2018) An exploratory study of Thai consumers' perceptions of "conspicuousness": a case of luxury handbags. J Consum Mark 35(6):601–612
7. Achabou MA, Dekhili S (2013) Luxury and sustainable development: Is there a match? J Bus Res 66(10):1896–1903
8. Griskevicius V, Tybur JM, Van den Bergh B (2010) Going green to be seen: Status, reputation, and conspicuous conservation. J Pers Soc Psychol 98(3):392–404
9. Amatulli C, De Angelis M, Korschun D, Romani S (2018) Consumers' perceptions of luxury brands' CSR initiatives: an investigation of the role of status and conspicuous consumption. J Clean Prod 194:277–287
10. De Angelis M, Adıgüzel F, Amatulli C (2017) The role of design similarity in consumers' evaluation of new green products: an investigation of luxury fashion brands. J Clean Prod 141:1515–1527
11. Janssen C, Vanhamme J, Lindgreen A, Lefebvre C (2014) The Catch-22 of responsible luxury: effects of luxury product characteristics on consumers' perception of fit with corporate social responsibility. J Bus Ethics 119(1):45–57
12. Athwal N, Wells VK, Carrigan M, Henninger CE (2019) Sustainable luxury marketing: a synthesis and research agenda. Inter J Manag Rev 21:405–426
13. Davies IA, Gutsche S (2016) Consumer motivations for mainstream "ethical" consumption. Eur J Mark 50(7/8):1326–1347
14. Wei CF, Chiang CT, Kou TC, Lee BC (2017) Toward sustainable livelihoods: Investigating the drivers of purchase behavior for green products. Bus Strateg Environ 26(5):626–639
15. Kapferer JN (1998) Why are we seduced by luxury brands? J Brand Manag 6(1):44–49
16. Nueno JL, Quelch JA (1998) The mass marketing of luxury. Bus Horiz 41(6):61–61
17. Lloret A (2016) Modeling corporate sustainability strategy. J Bus Res 69(2):418–425
18. Seo Y, Buchanan-Oliver M (2019) Constructing a typology of luxury brand practices. J Bus Res 99:414–421
19. Gardetti MA, Torres AL (eds) Sustainability in fashion and textiles: values, design, production and consumption. Routledge.
20. McNeill L, Moore R (2015) Sustainable fashion consumption and the fast fashion conundrum: fashionable consumers and attitudes to sustainability in clothing choice. Int J Consum Stud 39(3):212–222
21. Kapferer JN (2010) All that glitters is not green: the challenge of sustainable luxury. Eur Business Rev 40–45
22. Beckham D, Voyer BG (2014) Can sustainability be luxurious? A mixed-method investigation of implicit and explicit attitudes towards sustainable luxury consumption. ACR North Am Adv 24:245–250
23. Kapferer JN, Michaut-Denizeau A (2014) Is luxury compatible with sustainability? Luxury consumers' viewpoint. J Brand Manag 21(1):1–22
24. Latter C, Phau I, Marchegiani C (2010) The roles of consumers need for uniqueness and status consumption in haute couture luxury brands. J Glob Fash Market 1(4):206–214
25. Lee M, Bae J, Koo DM (2020) The effect of materialism on conspicuous vs inconspicuous luxury consumption: focused on need for uniqueness, self-monitoring and self-construal. Asia Pacific J Market Logist, in press
26. Shao W, Grace D, Ross M (2019) Consumer motivation and luxury consumption: testing moderating effects. J Retail Consum Serv 46:33–44
27. Vigneron F, Johnson LW (2004) Measuring perceptions of brand luxury. J Brand Manag 11(6):484–506
28. Guido G, Amatulli C, Peluso AM, De Matteis C, Piper L, Pino G (2020) Measuring internalized versus externalized luxury consumption motivations and consumers' segmentation. Italian J Market 1–23.
29. Lynn M, Harris J (1997) Individual differences in the pursuit of self-uniqueness through consumption. J Appl Soc Psychol 27(21):1861–1883

30. Brunner CB, Ullrich S, Jungen P, Esch FR (2016) Impact of symbolic product design on brand evaluations. J Product Brand Manag 25(3):307–320
31. Dubois B, Duquesne P (1993) The market for luxury goods: income versus culture. Eur J Mark 27(1):35–44
32. Sanford AJ, Moxey LM (2003) New perspectives on the expression of quantity. Curr Dir Psychol Sci 12(6):240–243
33. Teigen KH, Juanchich M, Riege AH (2013) Improbable outcomes: Infrequent or extraordinary? Cognition 127(1):119–139
34. Reich T, Kupor DM, Smith RK (2017) Made by mistake: when mistakes increase product preference. J Consumer Res 44(5):1085–1103
35. Bhattacharjee A, Mogilner C (2013) Happiness from ordinary and extraordinary experiences. J Consumer Res 41(1):1–17
36. Mason R (1984) Conspicuous consumption: a literature review. Eur J Mark 18(3):26–39
37. Bernheim BD (1994) A theory of conformity. J Polit Econ 102(5):841–877
38. Marcoux JS, Filiatrault P, Cheron E (1997) The attitudes underlying preferences of young urban educated Polish consumers towards products made in western countries. J Int Consum Mark 9(4):5–29
39. O'Cass A, Frost H (2002) Status brands: examining the effects of non-product-related brand associations on status and conspicuous consumption. J Product Brand Manag 11(2):67–88
40. O'Cass A, McEwen H (2004) Exploring consumer status and conspicuous consumption. J Consumer Behav: Int Res Rev 4(1):25–39
41. Jaikumar S, Sarin A (2015) Conspicuous consumption and income inequality in an emerging economy: evidence from India. Mark Lett 26(3):279–292
42. Wang Y, Griskevicius V (2014) Conspicuous consumption, relationships, and rivals: Women's luxury products as signals to other women. J Consumer Res 40(5):834–854
43. Chaudhuri H, Mazumdar S, Ghoshal A (2011) Conspicuous consumption orientation: conceptualisation, scale development and validation. J Consum Behav 10(4):216–224
44. Egea JMO, de Frutos NG (2013) Toward consumption reduction: an environmentally motivated perspective. Psychol Mark 30(8):660–675
45. Amatulli C, De Angelis M, Costabile M, Guido G (2017) Sustainable luxury brands: Evidence from research and implications for managers. Springer, Berlin
46. Eastman JK, Eastman KL (2015) Conceptualizing a model of status consumption theory: an exploration of the antecedents and consequences of the motivation to consume for status. Mark Manag J 25(1):1–15
47. Malhotra NK, Kim SS, Patil A (2006) Common method variance in IS research: a comparison of alternative approaches and a reanalysis of past research. Manage Sci 52(12):1865–1883
48. Podsakoff PM, MacKenzie SB, Lee JY, Podsakoff NP (2003) Common method biases in behavioral research: a critical review of the literature and recommended remedies. J Appl Psychol 88(5):879
49. Tehseen S, Ramayah T, Sajilan S (2017) Testing and controlling for common method variance: a review of available methods. J Manag Sci 4(2):142–168

Uzbekistan: The Silk Route of Handloom

Karan Khurana⬭

Abstract In the heart of central Asia lies the saga of handwoven silk and cotton in Uzbekistan. The textile sector of the nation is of great socio-economic importance since the Soviet era. The nation has exported silk and cotton across Central Asia and Russia for a long period. However, in this process of commercialization handloom lost its charm and market space. This article aims to revive the handloom sector back to the national and international fashion and textile trade. This research revolves around silk and cotton handloom production in the main regions of the country. The research is planned around the factories in the regions to figure out challenges and opportunities for the sector. To begin with, observational research will be held in the regions to see the presence and current status of the manufacturers. Secondary data was analyzed and qualitative interviews were conducted to analyze problems from the factory owner/ artisans. Lastly, solutions were provided in light of empirical evidence. This research significantly adds to the current literature as there is scarce data and researches in this sector. The handloom sector needs urgent attention as it is going through tough times. This article shall serve as a wake-up call for the stakeholders to initiate innovation and progress for a sustainable future. It shall also serve as an example for other regions in Uzbekistan and around Central Asia.

Keywords Sustainability · Handloom · Socio-economic development · Emerging economies · Cultural heritage

1 Introduction

Uzbekistan is a country of the ancient and original story, whose peoples have contributed much to world history. The territory of Uzbekistan is one of the sources of development of the original man (Shirinov et al. in Obshchestvennye nauki v Uzbekistane, no 8:57, 1993).

K. Khurana (✉)
School of Business and Economics (SOBE), Westminster International University, Tashkent, Uzbekistan

© The Author(s), under exclusive license to Springer Nature Singapore Pte Ltd. 2021   281
M. Á Gardetti and S. S. Muthu (eds.), *Handloom Sustainability and Culture*,
Sustainable Textiles: Production, Processing, Manufacturing & Chemistry,
https://doi.org/10.1007/978-981-16-5967-6_13

Uzbek culture is one of the novel and liveliest cultures of the east. The foundation of Uzbek culture dates back to the 6–seventh centuries B.C., as the nomadic tribes settled in the valleys of the Amu Darya, Syr Darya, and Zarafshan. These human settlements brought with them customs and traditions which eventually became the basis of Uzbek culture. Islam was introduced in modern Uzbekistan in the eighth century and has shaped every aspect of its culture since then.

From the famed blue-and-white ceramic pottery and detailed gold-thread embroidery to the intricately carved gourd snuffboxes, engraved copper lamps, and hand-forged daggers, Uzbek arts continue to enthrall a world audience. The fertile Fergana Valley is the most densely populated region in Central Asia (comprising Kazakhstan, Uzbekistan, Kyrgyzstan, Tajikistan, and Turkmenistan). This region is an intermontane depression nestled in the foothills of Tien Shan, up to 300 km in length from the West to the East and up to 170 km from north to south, is one of the main agricultural regions of Central Asia [5]. In the nineteenth century, many kinds of handicrafts developed in the Fergana Valley, such as silk materials, cotton textiles, paper, and pottery [24] which were the source of socio-economic development in the region. The "Queen of Textile"- Silk is also native and has provided worldwide recognition to the land. Sericulture dates back to the fourth century[1] in the Fergana valley, originating from the Zaravshan River and in the southern part of the country. The remnant trademark today is colorful silk ikat 'atlas' with arrow-featured splash patterns. Uzbek weaving masters in the Fergana Valley have developed a state-of-the-art technique of thread weaving, turning Margilan silk into one of the most desirable fabrics on earth (Picture 1).

Further on, the Suzani, a decorative and embroidered tribal textile material dates from the late nineteenth century Bukhara [18, 19]. The handwoven fabric for suzanis is known as kasbah, and it indicates the reputation of Suzanis in Uzbek society that there is a system and a word for the production of fabric [29]. Popular design motifs of the embroidery include sun and moon disks, flowers (especially tulips, carnations, and irises), leaves and vines, fruits (especially pomegranates), and occasional fish and birds.[2]

Uzbekistan is the sixth-largest cotton producer and the fifth-largest cotton exporter in the world 2.3 million tons of raw cotton in 2018,[3] exhibiting a well-developed structure concentrated on cotton. [12, 26]. Kawabata [20] reports that the country is ranked third after China and India in cocoon production and provides more than 80% of the cocoon production to the Commonwealth of Independent States (CIS). According to official figures, there are more than 2,000 companies in Uzbekistan working in textiles, providing jobs for at least 365,000 people.[4] It is evident that the textile sector has contributed to the socio-economic progress of the country, however, the handloom sector has not received much attention in the fashion markets. The academic literature in the past have highlighted various benefits for the emerging

---

[1] https://www.britannica.com/place/Uzbekistan/Economy.

[2] https://craftatlas.co/crafts/suzani.

[3] http://www.xinhuanet.com/english/2019-02/12/c_137815290.htm.

[4] https://eurasianet.org/uzbekistan-moves-to-lower-cost-of-cotton-for-manufacturers.

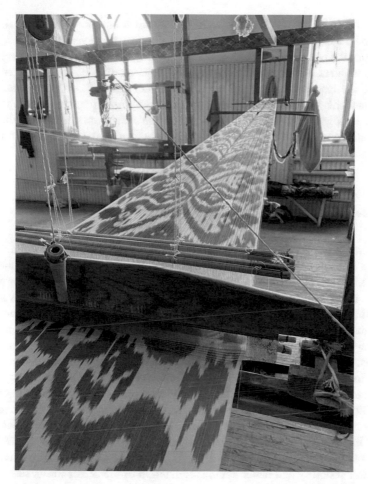

Picture 1 : Ikat Handloom. *Source* Author

economies such as the source of new entrepreneurship and income generation among young people [40], transforming the cultural identity of communities and countries by promoting competitiveness, creativity, design and innovation [31] and generating a high growth rate of gross domestic product [37, 43]. Handloom textiles are also attracting growing interest in fashion markets because of increasing concern about exploitation in production, thus encouraging interest in the economic benefits of fairly traded, high-quality materials and the potential contribution of handloom to sustainability in the fashion industry [46]. Besides commercial advantages, cultural heritage embodied in traditional crafts is an integral part of any nation which reflects the culture and tradition of a particular region [47]. In emerging economies, industrialization and globalization have challenged the survival of artisans and gradually have led to the extinction of handlooms. The existence of handloom in fashion markets is an answer to the mass production and monotony of the clothing design that we observe today. In an era of globalization and rapid economic change, this heritage needs to be identified and protected or it may disappear forever [14, 47].

Despite the benefits of the handloom to emerging economies, it was observed that the stakeholders of the value chain suffer various challenges. This chapter brings focus to the Handloom sector of the nation which has been overshadowed by the cotton and silk textile exports. On doing secondary research, the authors find that there is no significant scientific literature on the handloom sector in Uzbekistan. The hindrances of the shall be analyzed and solutions shall be provided in light of empirical evidence from the author.

2 The Challenges of the Handloom Sector

Uzbekistan's geographic location and tourism resources have made it a favorite amongst travelers. A rich cultural and historical heritage, location along the Great Silk Road, as well as ancient traditions and in culture and arts can attract to the country cultural, religious, archaeological, and ethnographical tourists [4]. The government has put in substantial efforts since the independence in 1991 to boot the potential and realize the industry to be a major socio-economic booster. However, the last couple of years have posed serious challenges across the handloom value chain. In the Covid-19 crisis, the handloom sector is withering as there is no tourism and citizens do not buy the local fabric or products. The pandemic has added a series of challenges to the sector and the sector needs major managerial interventions. This section delves into the challenges faced by the stakeholders today. The handloom and textile industry have been a socio-economic catalyst to the economy since the ex-Soviet Union era but has not achieved the desired progress ever since. To avoid the extinction of this historical craft, this article aims to provide a set of managerial solutions for greater visibility and better commercialization of the sector for its stakeholders (suppliers, domestic brands, manufacturing companies, foreign investors, academia, governmental bodies, and consumers). It shall ensure holistic development of the artisan communities for a sustainable future.

2.1 Raw Material Availability and Increasing Prices

Until today the common trend has been to export cotton to other countries for an increased trade potential in the world market. Uzbekistan's exports of cotton yarn, textiles, and readymade garments were about US$1.1 billion in 2016 and were estimated to reach US$1.3 billion at the end of 2017. Iran, China, Bangladesh, Turkey, Russia, Latvia, Hong Kong, and CIS countries were the main buyers for Uzbek cotton in 2018. In 2018, 115.6 thousand tons of cotton fiber were reportedly exported totaling worth US $ 222.1 million.[5] The Ministry of Agriculture announced a payment of $0.10 to $0.40 for every kilogram of hand-harvested cotton, marking a significant increase from the 2019 rates of $0.02 to $0.08 per kilogram. The Ministry also established a minimum price for raw cotton harvested by machine. The new price is 920,000 Uzbek som per ton, or approximately $89, compared with a price of 600,000 Uzbek soum, approximately $63 at the time, per ton in 2019.[6] This price increase has to fight the forced labor in the fields and the current COVID-10 crisis in the textile trade. However, it makes raw material less available and costly for the weavers across the country. This is one of the problems which was highlighted by the manufacturers that makes it difficult to sustain production and bring more innovative products to the market (Picture 2).

2.2 Financial Constraints

Lack of financial resources has been a classic pressing issue for the handloom sector across the world. Various authors [1, 9, 32] have highlighted that shortage of capital, and low-profit income is the key factors for the decline of handloom weaving ventures. In the discussion with the artisans, it was found that they have negligible support from the government authorities in terms of financial planning and management. Their cash flow was based on the sales which were regional and due to the drop in national tourism they had no cash flow in the companies. This also highlights a major disruption in the handloom value chain in the country where the artisan in regions is not exposed to various other channels of retailing and is heavily dependent on the tourism sector.

2.3 Lack of Modern Technology

The handloom industry in the republic is decentralized, household-based, and based on labor inputs from the household members. While visiting the factories and work-shops it was realized that the looms and old fashioned and it is not practically possible to achieve higher productivity. Bhavani [6] argues that the poor adoption of technology in the industry hampers the expansion of the handloom trade and forced the industry to depend heavily on human laborers, who carry out their work either at home, the typical cottage industry, or at small and medium-sized enterprises. Further, the technologies used are traditional and affect the quality of the final product (see

---

[5] https://tashkenttimes.uz/economy/3432-uzbekistan-considerably-decreases-cotton-fiber-exports-in-2018.

[6] https://www.caspianpolicy.org/uzbekistan-implements-changes-in-cotton-industry-to-stimulate-economy/#:~:text=The%20Ministry%20of%20Agriculture%20announced,raw%20cotton%20harvested%20by%20machine.

Picture 2 Raw cotton for handloom. *Source* Author

Picture 3). This has a long-lasting impact on the consumers as they do not consider the handloom to be a good quality product (Picture 3).

2.4 Lack of Merchandise Innovation

Art and craft bazaars in the capital city are one of the main retailing channels in the country for the handlooms sector. Merchandise such as scarves, carpets, pillow covers, traditional dresses, and embroidered fabrics are displayed for the public and tourists. For this research, the author visited three locations (Art center, Next mall, and International hotel) for over six months to see the variety of merchandise available. It was found that the merchandise was similar in all the period. There were no new designs or products that could attract the consumers and make them a re-purchase of the products.

This exhibits a huge gap in innovation in the sector which has led to the loss of interest from the domestic consumer. Authors [2, 16] state that design and innovation could easily turn the cultural heritage of handlooms into the commercial industry and profit the stakeholders. Hence, there is an immense need for novel merchandise to attract consumers (Picture 4).

Picture 3 Raw material dying. *Source* Author

Picture 4 Ikat Handloom designs

2.5   Lack of Data on Craftsmen

During this research, one of the most noteworthy problems is that there is no documentation on the craftsmen in the country. It was almost impossible to find data on even highly appraised handicrafts of the nation. This problem disconnects the artisans from the international development cooperations working in the area of arts and culture. These organizations can help them to develop and solve various problems that are hindering the growth of the sector. There lies an urgent need to document the clusters and the artisans employed in the sector so that timely solutions could be provided.

3   Recommendations and Discussions

From Sect. 2, it is evident that the value chain suffers from managerial as well as production problems. Local entrepreneurs/ communities or artisans (flagbearers of culture and heritage) of Uzbekistan produce traditional crafts (embroidery, hand weaving, Ikat silk handlooms, etc.) are classified as SMEs. In Uzbekistan, the share of SMEs' exports in total country export performance increased to 27.2% in 2017 from 6% in 2005 [42] and hereby it becomes very critical to save the sector. In this section, the author recommends a viable business model following the empirical evidence for the SME's in Uzbekistan.

3.1   Ensuring Raw Material Supply

The availability and timely delivery of raw materials are some of the key areas to begin a strong supply chain. In handloom economies (India and Sri Lanka) it was found that a constant and cheap raw material supply was very influential in the success of the SME's [36]. In the case of Uzbekistan, the raw material (cotton and silk) are available in abundance. However, rising prices and concentration on exports have often left the artisans in inventory shortage. The government plays a very vital role in facilitating the raw material supply to the artisans. For instance, in India, the government issued a Yarn Supply Scheme (YSS) which was implemented through National Handloom Development Corporation. Under this scheme, freight is reimbursed and depot operating charges at 2%, and a component of 10% price subsidy also exists on hank yarn, which is applicable on cotton, domestic silk, and woolen yarn with quantity caps.

Implementation of such schemes has two major advantages; first that it ensures a constant supply of raw materials to the handloom SME's and second that it documents the number of artisans employed in the sector which is currently missing. Existing government bodies Uzbekipaksanoat[7] and Uztekstilprom[8] should look into making such schemes to uplift the artisans and entrepreneurs in the handloom sector.

3.2 Innovation as a Strategy

In the current scenario, where fast fashion has made clothing mundane, the handloom products exhibit an exciting pathway to premium and luxury segments. The uniqueness and splash of historic art in textiles is a way forward in luxury [8, 17]. However, the handloom products lack innovation, novelty, and quality. Prathap and Sreelaksmi [38] state that quality consciousness, perceived quality, and product trust have a significant influence on consumers' decision to purchase traditional handloom products. Mamidipudi [30] show how weaving communities in India are constantly innovating their technologies, designs, and social organization to remain abreast with the consumer demands. The Uzbek handloom products haven't changed for ages and as a result, it is found out of trend. The current merchandise is very limited as seen in bazaars and does not gain consumer interest. However, secondary data shows that designers from Europe and U.S.A sourced the Uzbek handloom fabric and made premium-priced products for their markets.

Handloom is a mechanical process and therefore, the innovation should take place in the fabric production stage. It should be noted here that it is not suggested to change the art form or lose the historic process. New designs and motifs could be created to produce a new range of fabrics. It can be agreed that the artisans might not be able to create and develop fashion merchandise as it is a highly skilled job and they do not specialize in that area. Taking into consideration the current status of the Uzbek fashion market, collaborations with existing Uzbek designers would be a very innovative notion. Such collaboration can bring the Uzbek local product to establish a mark in the global market. Nevertheless, this collaboration should not overshadow the creativity of the artisan. Alongside a fair pay structure, the government authorities should ensure the protection of intellectual property rights for the artisans. In such a way a large merchandise assortment can be made and presented to the local and international consumer.

3.3 The Digital Connection

The handloom SMEs have almost no footprint on the digital space. A few brands are present on Etsy and some travel blogs talk about the handloom and crafts of Uzbekistan. It is practically impossible to find an artisan who specializes in a certain kind of handloom style. The COVID-19 crisis has mandated a digital presence for

---

[7] https://www.uzbekipaksanoat.uz/en/.

[8] https://invest.gov.uz/partnery/native-organizations/uztekstilprom/.

fashion and textile companies. Companies are beginning to realize that the market has changed and that consumers are by nature omnichannel, i.e. they access the company through different channels [28].

Digital promotion and selling will serve various advantages to the artisan communities:

- E-commerce and M-commerce serve as a longer-term strategy to sell merchandise in Central Asia and beyond. This shall reduce their physical retailing costs (exhibition /shops) and finally increase profit margins.
- A digital trade path shall remove the historic middle man problem that has existed for ages [15, 39].
- Promotion on various social media platforms shall increase consumer awareness and interest in the handloom product. Past researches [7, 27] have proven that consumer is highly influenced by electronic word-of-mouth (e-WOM).
- Lastly, it shall also be a shock absorber against future economic crises.

In May 2018, the Uzbek President signed a decree "On measures for the accelerated development of e-commerce", and approved the "Program for the Development of E-commerce in Uzbekistan for 2018–2021". Digitization has emerged as the "new normal" and the artisans need to promote through social media such as Instagram, telegram, and Facebook. Basic knowledge on how to use social media can make the artisan visible in the digital space. Currently, there are a few domestic online selling platforms (Express24, Bulavka. uz) that could help increase the sales of handloom in the country. An omnichannel strategy shall decrease the gap between the buyer and seller and finally result in a much-required supply chain redesign [35, 45].

3.4 Financial Modelling and Governance

In emerging economies, financial governance for the handloom sector has always been a pestering problem. Academic literature [21, 23, 25] has shown that efficient governance is a major requirement for the survival of the arts and culture sector in emerging economies. The ministry of Textiles should take special attention to directing the funds through controlled schemes so that the SME's could be benefited. A good example of such governance can be drawn from the Indian handloom sector. To begin with, the isolated artisans and SME's should be clustered and the financial hindrances should be studied in each cluster. Then, a special bank (such as National Bank for Agriculture and Rural Development (NABARD)[9] in India) should be implemented that handles the schemes and quotas for the handloom sector. Examples of strategic activities of such an organization could be strengthening of weaver clusters, credit guarantee for fresh loans, loss assessment exercise, waiver of loans, and recapitalization of individual weavers and artisans. Lastly, these clusters should be monitored on a periodic and continuous basis. For this Uztekstilprom and

---

[9] https://www.nabard.org/default.aspx.

Uzbekipaksanoat could assign departments that shall specially register and deal only with the handloom sector. They shall also hold the responsibility to promote the final products to world markets as a part of cultural heritage. The combination and financial modeling and governance shall be an effective step to structure the sector and bring in notable results in the future.

3.5 Supporting Sustainable Development and Training Human Resources

The handloom sector is a socio-economic booster to developing nations as it targets to develop the disadvantaged sections of the society [10, 13] and contributes to GDP contributions, foreign exchange, and employment generation [33]. The second view of Handloom is from the perspective of sustainability. It consumes minimal power and eco-friendly and possesses all the capability to grow and innovate [11]. And finally, handloom serves as country's souvenirs for tourists.

To benefit from this sector the human resource should receive a structured human resource development. Ogbeibu et al. [34] argue that green human resource management (GHRM) practices such as green training, green recruitment and green performance measures enhance workers' involvement in green activities. Secondly, a very significant area to concentrate on is the women involved in this sector. Research shows that women in the sector face several demographic and structural challenges that prevent them from effectively engaging in the economy and employment outside the home. According to the world bank report around 60 percent of rural women have received only a general secondary education (grades 9 to 11) or below, and only around 8 percent have obtained higher education. The apparel producing economies (such as India and Bangladesh) have constituted ITI (Industrial training institutes) and TVET (Technical and Vocational Education and Training) to generate qualified human resources to fill in the value chain and subsequently solving the problem of jobless citizens. These training institutes shall be very instrumental in raising the skill level of artisans and increasing productivity which is currently missing in Uzbekistan.

4 Conclusions

Uzbekistan set an ambitious goal to achieve the status of an upper-middle-income country by 2030 with the focus on growing a large and solid middle-class social stratum. The current reforms in Uzbekistan largely target the development of small and medium-scale businesses and entrepreneurship. These developments are the primary solution to acute social problems such as unemployment, poverty — especially among women and youth — and poor quality of life [42]. Each industry shall have a noteworthy contribution to the nation's progress to make a mark on world trade. The nation is taking active measures to develop the tourism industry to reap its benefits as foreign visitor inflow into Uzbekistan has grown from 300,000 in 2000 to more than 2 m in 2016, a sevenfold increase [44], Allaberganov and [3]. Local crafts are important elements of heritage and culture, and people travel to see and experience such unique creations of mankind. Moreover, the handloom of Uzbekistan shall help the nation to make a mark in the fashion sphere. The domestic fashion consumer market is stocked with Chinese and Turkish merchandise and has captured the market share of domestic and traditional Uzbek clothing. The existing products are of very low quality and style and hereby do not match consumer preferences,as a result, the millennial and fashion-conscious consumers are spending all their disposable income on imports which is a huge loss to the country's economic environment [22].

Currently, there is no academic literature on Uzbek handloom. Through this article, the authors have tried to trigger the thoughts of the stakeholders in the Uzbek textile value chain. It is about time to think in a holistic way for the preservation of the Uzbek handloom otherwise, the craft might lead to extinction. The handloom sector has suffered a lot during the covid-19 crisis and has led the artisans to a very difficult stage. This article shall serve as a wake-up call for the stakeholders to initiate innovation and progress for a sustainable future. It shall also serve as an example for other countries of Central Asia.

5 Further Scope of Research

Current research studies focus on monuments and other historic elements of Uzbek culture. However, through this research, it is evident that handloom requires immediate attention across the entire value chain. Further studies can be conducted to re-engineer each stage of the value chain and make the sector profitable. Educational and governmental organizations can take up research projects which shall be an excellent connection between academia, handloom, and the government sector. Holistically, this shall also encourage youth entrepreneurship amongst various university students which are currently absent in the sector. Lastly, as handloom is connected to tourism sector which is a top priority of the government today and hence, many simultaneous researches could be performed that benefit both sectors.

References

1. Ahmed Z, Hussain AB, Alam R, Singha AK (2021) Perils and prospects of Manipuri handloom industries in Bangladesh: an ethnic community development perspective. GeoJournal 1–16
2. Anurag P, Das K (2020) Inclusion and innovation challenges in handloom clusters of Assam. In Inclusive Innovation, Springer, New Delhi, pp. 75–99

3. Allaberganov A, Preko A (2021) Inbound international tourists' demographics and travel motives: views from Uzbekistan. J Hosp Tour Insights. Vol. ahead-of-print no. ahead-of-print. https://doi.org/10.1108/JHTI-09-2020-0181

4. Aleksandra K (2013) Analysis and perspectives of tourism development in Uzbekistan. AGALI J Soc Sci HumIties 3(3):87–101

5. Artem D (2007) Ferghana valley: problems of maintaining economic stability. Central Asia and the Caucasus 2(44)

6. Bhavani TA (2002) Small-scale units in the era of globalisation: problems and prospects. Econ Polit Wkly, 3041–3052

7. Chu SC, Kim Y (2011) Determinants of consumer engagement in electronic word-of-mouth (eWOM) in social networking sites. Int J Advert 30(1):47–75

8. Clifford R (2018) Learning to weave for the luxury Indian and global fashion industries: the handloom school. Maheshwar. Clothing Cultures 5(1):111–130

9. Dash M (2020) Emerging entrepreneurship opportunities in handloom sector with special reference to Odisha. Psychol Educ J 57(9):783–791

10. Dasgupta A, Chandra B (2016) Evolving motives for fair trade consumption: a qualitative study on handicraft consumers of India. The Anthropologist 23(3):414–422

11. Dissanayake DGK, Perera S, Wanniarachchi T (2017) Sustainable and ethical manufacturing: a case study from handloom industry. TextEs Cloth Sustain 3(1):1–10

12. Djanibekov N, Rudenko I, Lamers J, Bobojonov I (2010) Case Study #7–9 pros and cons of cotton production in Uzbekistan. In: Andersen Per P, Cheng F (eds). food policy for developing countries: case studies. pp. 13

13. Forero-Montaña J, Zimmerman JK, Santiago LE (2018) Analysis of the potential of small-scale enterprises of artisans and sawyers as instruments for sustainable forest management in Puerto Rico. J Sustain For 37(3):257–269

14. Hani U, Azzadina I, Sianipar CPM, Setyagung EH, Ishii T (2012) Preserving cultural heritage through creative industry: a lesson from Saung Angklung Udjo. Procedia Econ Financ 4:193–200

15. Hani U, Das A (2017) Design intervention in the handloom industry of Assam: in the context of a debate between traditional and contemporary practice. In: International Conference on Research into Design, Springer, Singapore, pp. 999–1006

16. Hani U, Das AK (2019) Eri-culture: the drive from tradition to innovation. In: research into design for a connected world. Springer, Singapore, pp. 835–845

17. Hitchcock L (2016) A lack of luxury? contemporary luxury fashion in Sri Lanka. Luxury 3(1–2):63–82

18. Izrailova A (1998) Suzani Vernacular: technique and design in the Central Asian dowry embroideries. In: Textile society of America symposium proceedings. 177. https://digitalcommons.unl.edu/tsaconf/177

19. Kalter J, Pavaloi M (eds) (1997) Uzbekistan: heirs to the silk road. Thames and Hudson

20. Kawabata Y (2016) Important structures of sericulture for world strategy in Uzbekistan (No. 923–2016–72931)

21. Khurana K (2019) Ethiopia. In cultural governance in a global context (pp. 21–50). Palgrave Macmillan, Cham. https://doi.org/10.1007/978-3-319-98860-3_2

22. Khurana K, Ataniyazova Z (2020) Insights and future forward for fashion and textile value chain in Uzbekistan. Res J Text Appar 24(4):389–408. https://doi.org/10.1108/RJTA-03-2020-0020

23. Khurana K, Saraceno M (2019) The current standing and future prospects of arts and culture in ethiopia. EUREKA: social and humanities, (4):3–15

24. Kikuta H (2009) A master is greater than a father: rearrangements of traditions among muslim artisans in Soviet and post-Soviet Uzbekistan. Economic development, integration, and morality in Asia and the Americas (Research in Economic Anthropology, Volume 29). Emerald Group Publishing Limited, 89–122

25. King IW, Schramme A (Eds.) (2019) Cultural governance in a global context: an international perspective on art organizations. Springer

26. Kim YJ, Park J (2019) A sustainable development strategy for the Uzbekistan textile industry: the results of a SWOT-AHP analysis. Sustain 11(17):4613
27. Litvin SW, Goldsmith RE, Pan B (2008) Electronic word-of-mouth in hospitality and tourism management. Tour Manage 29(3):458–468
28. Lorenzo-Romero C, Andrés-Martínez ME, Mondéjar-Jiménez JA (2020) Omnichannel in the fashion industry: a qualitative analysis from a supply-side perspective. Heliyon, 6(6), e04198
29. Ludington S (2018) Embroidering paradise: Suzanis as a place of creative agency and acculturation for Uzbek women in 19th century Bukhara In: Textile society of America symposium proceedings. 1094. https://digitalcommons.unl.edu/tsaconf/1094
30. Mamidipudi A, Bijker WE (2018) Innovation in Indian handloom weaving. Technol Cult 59(3):509–545
31. Moalosi R, Popovic V, Hickling-Hudson A (2010) Culture-orientated product design. Int J Technol Des Educ 20(2):175–190
32. Naidu S, Chand A (2012) A comparative study of the financial problems faced by micro, small and medium enterprises in the manufacturing sector of Fiji and Tonga. Int J Emerg Mark 7(3):245
33. Naidu S, Chand A, Southgate P (2014) Determinants of innovation in the handicraft industry of Fiji and Tonga: an empirical analysis from a tourism perspective. J Enterprising Communities 8(4):318
34. Ogbeibu S, Emelifeonwu J, Senadjki A, Gaskin J, Kaivo-oja J (2020) Technological turbulence and greening of team creativity, product innovation, and human resource management: Implications for sustainability. J Clean Prod, 244, 118703
35. Piotrowicz W, Cuthbertson R (2014) Introduction to the special issue information technology in retail: toward omnichannel retailing. Int J Electron Commer 18(4):5–16
36. Prasad S, Tata J (2010) Micro-enterprise supply chain management in developing countries. J Adv Manag Res 7(1):8
37. Raihan MA (2010) Handloom: an option to fight rural poverty in Bangladesh. Asia-Pac J Rural Dev 20(1):113–130
38. Prathap SK, Sreelaksmi CC (2020) Determinants of purchase intention of traditional handloom apparels with geographical indication among Indian consumers. J HumIties Appl Soc Sci Vol. ahead-of-print No. ahead-of-print. https://doi.org/10.1108/JHASS-04-2020-0055
39. Samira P, Dey SK (2015) Profitability analysis of handloom weavers: a case study of Cuttack district of Odisha. Abhinav-National Monthly Refereed J Res Commer Manag 4(8):11–19
40. Setyaningsih S, Rucita CP, Hani U, Rachmania IN (2012) Women empowerment through creative industry: a case study. Procedia Econ Financ 4:213–222
41. Shirinov T, Anarbaev AA, Buriakov YF (1993) Arkheologicheskie issledovaniia v ANRU. Obshchestvennye nauki v Uzbekistane, no 8:57
42. Tadjibaeva D (2019) Small and medium-sized enterprise finance in Uzbekistan: challenges and opportunities. ADBI working paper 997. Tokyo: Asian development bank institute. Available: https://www.adb.org/publications/small-medium-sized-enterprise-finance-uzb ekistan-challenges-opportunities
43. Usero B, Angel del Brío J (2011) Review of the 2009 UNESCO framework for cultural statistics. Cult Trends 20(2) 193–197
44. Uzstat (2017) Main indicators of recreation and tourism development in the Republic of Uzbekistan. Available at: https://stat.uz/en/435-analiticheskie-materialy-en1/2062-main-indicators-ofrecreation-and-tourism-development-in-the-republic-of-uzbekistan
45. Verhoef PC, Kannan PK, Inman JJ (2015) From multi-channel retailing to omni-channel retailing: introduction to the special issue on multi-channel retailing. J Retail 91(2):174–181
46. Wanniarachchi T, Dissanayake K, Downs C (2020) Improving sustainability and encouraging innovation in traditional craft sectors: the case of the Sri Lankan handloom industry. Res J Text Appar 24(2):111–130. https://doi.org/10.1108/RJTA-09-2019-0041
47. Yang Y, Shafi M, Song X, Yang R (2018) Preservation of cultural heritage embodied in traditional crafts in the developing countries. a case study of Pakistani handicraft industry. Sustain 10(5) 1336

Aesthetic Capitalism and Sustainable Competitiveness in Urban Artisanal Networks

WenYing Claire Shih and Konstantinos Agrafiotis

Abstract The UN's Agenda 2030 refers to Sustainable Development Goals (SDGs) as environmental conditions are rapidly deteriorating. The goals serve as a guide in order to improve the human condition in the future. The UN also has promoted the use of the creative economy to generate economic value, wealth and jobs. The so-called aestheticization of commerce refers to a condition where the world has been introduced to 'Aesthetic Capitalism' where a new concept of aesthetics serves as an antidote to the boredom which has afflicted consumers in contemporary societies. Aesthetics have become a mechanism to generate renewed interest in products and services where fascination is the dominant concept in this new aspect of capitalism. The authors attempt to explore the ethics of sustainable development in relation to those of aesthetics in its capitalist context and also explore empirically the emergence of a counter current to the luxury world, detached from the ubiquitous luxury brands and their global domination. This alternative view of luxury is more associated with local production and consumption where craft, personalization, empathy and ethics have a dominant role. The questions raised pertain to how aesthetic capitalism can also be ethical, how it can fit into this alternative notion of luxury, and how in urban artisanal networks, micro fashion businesses can remain viable and competitive in the medium term and contribute to the city's quality of life and inclusive growth. Research methodology follows the interpretivist theory integral to the qualitative research tradition. A single case study has been conducted in Taiwan which points to the idea that aesthetic capitalism can be symbiotic with urban artisanal luxury where crafters endeavour to satisfy customers' pursuit of hand-crafted artefacts, in a manner detached from expensive branding and conspicuous consumption. This admittedly small subsector of craft-oriented luxury may be on the rise as it lends support to the UN's concepts of sustainable development and improvement of the city's quality of life.

W. Claire Shih (✉)
Department of Innovative Living Design, Oversea Chinese University, 100 Chiao Kwang Rd, Taichung, 40721 Taichung City, Taiwan
e-mail: wycshih@ocu.edu.tw

K. Agrafiotis
Independent Researcher and Consultant in the Clothing Sector, Thessaloniki, Greece

295
M. Á Gardetti and S. S. Muthu (eds.), *Handloom Sustainability and Culture*, Sustainable Textiles: Production, Processing, Manufacturing & Chemistry, https://doi.org/10.1007/978-981-16-5967-6_14

Keywords Ethics and aesthetics · Sustainable development · Urban artisanal
networks · Taiwan

1 Introduction and Background

The sustainable development goals promulgated by the UN refer to the so-called
Agenda 2030 where UN member states agreed on 17 goals and 169 associated
targets for the overall betterment of the human condition [31]. Three sustainable
development goals (SDGs) relate more specifically to this study. These include SDG
8 which refers to decent work and inclusive sustainable growth; SDG 12 referring
to sustainable production and consumption, where economic growth is detached
from environmental degradation, thus increasing resource efficiency and promoting
sustainable lifestyles; and SDG 11 referring to sustainable cities. In a related area, the
Intergovernmental Science Policy on Biodiversity and Ecosystem Services (IPBES)
has reported in a "Summary for Policymakers" also supported by the UN, explicitly
stating that sustaining the planet's living systems demands that humanity needs to
redefine the meaning of the term "quality of life" as studies show that happiness is
connected more to lowering consumption and reducing waste. Therefore, humanity
needs to divert its attention from the current mind-set that a good and meaningful
life is related mainly to excessive material consumption. This rejects the notion of
GDP as the only important metric, but other metrics should be included, citizens'
well-being the principal one [11].

Another concept also introduced by the UN is the creative economy. This has
become a significant and fast growing economic sector as it generates jobs and creates
wealth by capturing economic value from the cultural expression and creative output
generated by the so-called creative class [10, 30]. The creative industries, which form
the focus of this study, combine expressive values with functional elements where
industrial design (the fashion industry in our case) is included in this subset. Since
the creative industries' output is designed to capture economic value through creative
expression, the concept of the aesthetization of commerce takes centre stage in the
discourse of creative value chains in contemporary societies [10].

In the authors' previous research projects, it has been argued that production
and consumption can occur within the urban setting where fashion-related artisans
produce small batches of highly specialized fashion artefacts while consumption
occurs in small shops, as both activities are located within the city. It has also been
argued that within the context of an urban economy, micro artisanal businesses organized as a network can thrive effectively in a less aggressive form than constant
competition. In this artisanal setting, discerning customers' concerns are centred
on knowing who made the fashion artefacts, in what place specificities and under
what conditions. Artisanal modes of production have also given rise to a renewed
interest in traditional skills and age-old crafts which were nearly extinct [17]. This
arrangement of production and consumption has been described by the authors as
urban artisanal luxury. Urban artisanal luxury relies more on local manufacture,

as by nature these fashion artefacts connect more with human emotions and with customers who wish to experience and enjoy the craft processes performed by local crafters [17, 26]. It relies less on branding, logos, extravagant retail environments and unjustifiably exorbitant prices which form the de-facto principles of the luxury industry [28]. To the best of the authors' knowledge, urban artisanal luxury has been a largely unexplored field of research inquiry which corresponds closer to the UN's SDGs.

In this chapter, the authors explore empirically how aesthetic capitalism fits in with the sustainable development goals promulgated by the UN; and also how in the fashion industry's subsector of urban artisanal networks, economic value is created by ethical local production and responsible consumption. In order to understand the relationships between aesthetic capitalism and sustainable development, the authors have explored how a micro fashion firm operates within an urban artisanal luxury network in Taiwan; how it has managed to remain viable and competitive; how realistically it has implemented ethical sustainable production and promoted responsible consumption; and how it has contributed to inclusive growth and quality of life within the urban setting. The theoretical framework comprises the concept of aesthetic capitalism, the relational view of the firm, urbanization economics and network resources, craft practice and the transference of tacit knowledge, slow fashion, and the true meaning of luxury and experiential retailing. At the end of the chapter, a conceptual framework is devised where the apparently contradictory forces of aesthetic capitalism and ethical and slow production and consumption are combined by forming an artisanal fashion network.

2 Literature Review

2.1 Aesthetic Capitalism and the Culture of Fashion Markets

In traditional capitalism, the formation of economic cycles, coincide with activities performed in tandem with stylistic and affective cycles where creativity and innovation take centre stage [10]. Therefore, the aestheticization of capitalism has affected most of contemporary societies. It is also argued that people in post-modern societies seem to be weary of consumption symbols and as a result they are reluctant to engage with this form of consumption [19].

This brings forth the concept of aesthetic capitalism which runs against the Keynesian economic model this of demand creating supply. In aesthetic capitalism the opposite emerges where supply rather generates demand [19]. Murphy et al. [19] argue that aesthetic capitalism has been manifesting itself in the supply part of the value chain which became more significant than manufacture and logistics. The reason for this is that style and image have become the prevailing parameters in this economic formation. Therefore, in current terms, the *supply of fascination* has prevailed where the crucial part of fascination hinges upon the fact that aesthetics can be captivating where function follows form and not the other way round as in Modernism principles [19], p. 8). This brings forward the notion of a certain metropolitan lifestyle which largely refers to aesthetic artefacts production which involve design, marketing and retailing as well as distribution media in the form of lifestyle magazines, digital platforms and so on which are concentrated on the metropolitan aspect of consumption. Aesthetic considerations here are of essence because aesthetics imbue artefacts with fascination, interest and desirability by adding interest value to the equation ([19], p. 8).

2.1.1 The Culture of Markets

Markets within the context of aesthetic capitalism contain a structural element which can be captured by looking at alliances formed between companies and collaborative teams of professionals or mobility among these professionals across companies [10]. Moreover, this social element in markets can also have a cultural dimension. This is called the 'culture of markets' as it can refer either to the culture of market stakeholders in terms of how they are socialized and what their beliefs and values are, or to the cultural aspects of creatives' output ([10], p. 73). Output here means design or media narratives which are crucial because they form the interface between the creatives and their audiences who evaluate them. These cultural aspects are particularly salient in creative industries such as fashion, film, and music where the production of so-called cultural products proliferates [10]. Therefore, customers buy cultural products not only for their use, value or function, but mostly for the meanings attached to them. The fashion industry can be a significant setting for understanding how social relations and meanings are intertwined. Moreover, fashion companies make extensive use of stylistic elements to present collections that will sell only if they are meaningful to customers. This does not imply that fashion products do not have functional uses, instead, this utilitarian element is supplanted by meaning. Therefore, customers usually base their purchasing decisions on aesthetic and stylistic cues [10].

Aesthetic capitalism refers to companies of various sizes operating within a cultural ecosystem within an urban setting. In the next section, the fundamental conditions by which firms can reach competitive advantage through the formation of networks is concisely discussed.

2.2 The Relational View of the Firm, Network Resources and Urbanization Economies

The relational view of the firm states that the company can be viewed as a nexus of relations where it may extend its boundaries and form relationships with other network members by accessing network resources. Thus, competitive advantage can be achieved within a network of inter-firm collaborations [7, 15, 16]. The extension of a company to others means that the company may operate within a specific geographical space. Territorial embeddedness, namely, short distances can be advantageous since it facilitates planned and also face-to-face interactions [15]. In relational networks, participant companies can have access to and simultaneously pool resources in order to generate competitiveness for themselves and the network in a cooperative manner [4, 17]. Therefore, networks appear to be more suitable for the present research than the industrial cluster theory [22]. In clusters, the dominant logic revolves around aggressive competition and rivalry in all actors' transactions [16]. Moreover, productivity rises is not the only metric for regional competitive advantage [22] because other parameters come to the fore as well such as environmental concerns and social cohesion in the distribution of economic progress [2].

These micro businesses organized as a network have more in common with the notion of cottage industries producing mostly hand-made artefacts. In this configuration, productivity rises and aggressive competition seem less important. Other important parameters of urban economies are also omitted from the clusters literature including employment in local production and retail which brings prosperity to locals and the presence of an urban creative class which can drive creative industries [14]. In the same line of argument, sustainable competitiveness currently features highly on governments' agendas with regards to environmental and social policy decisions because it promotes higher levels of sustainable prosperity [2]. Thus, city councils and a new generation of entrepreneurs constantly seek new arrangements in order to grow and sustain prosperity in cities and regions as studies suggest that inclusive growth (high levels of community employment) is directly related to competitiveness [2, 5]. This introduces the concept of urbanization economies where a diverse number of industries are concentrated within the city as this characteristic generates opportunities for the cross-fertilization of ideas in related urban industries.

Dynamic urban economies usually hinge upon a particular form of organizing businesses, a skilled labour pool, specialized suppliers and the diffusion of knowledge in a variety of spillovers [5, 25]. Employment becomes a significant factor in the urban economy equation because it refers to peoples' well-being in the area. Kitson et al. [14] propose an alternative model of competitiveness to Porter's diamond for regional economies which has been adapted for this chapter. In this, the quality and skills of the labour force, the sociality of networks in either physical or digital form, the range and quality of cultural institutions and infrastructure coupled with a creative or innovative class can lead to improved levels in the quality of life and inclusive growth. It is acknowledged that Porter's contribution to this arrangement is *demand conditions* as these can make or break any network because customer sophistication is paramount as it can drive sales and support financially the designers and crafter network, thus contributing to its survival, competitiveness and reputation potential (see Fig. 1).

In the next section, the slow fashion principles, craft and tacit knowledge transfer are discussed as these form essential values in small scale production and consumption performed by micro-fashion businesses within the urban setting.

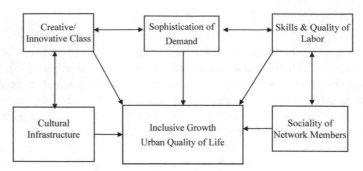

Fig. 1 Urbanization Economies. *Source* Adapted by the authors

2.3 Slow Fashion, Craft and Tacit Knowledge Transfer and Sharing

Slow fashion engages people to think in a more responsible manner about the concepts of sustainable production and consumption as it requires potential customers to acquire more detailed knowledge of a slow fashion product before its purchase. Fletcher [8] in her seminal research of slow textiles and fashion sets out a number of fundamental principles. Firstly, the customer becomes co-producer as he/she participates in the production process. The customer co-producer also prefers quality to quantity. Secondly, companies within the slow fashion mind-set engage in fair practices towards workers across the supply chain, thus improving their livelihoods. Thirdly, companies' managers become more resourceful in alternative ways of achieving natural resource preservation. They use mostly local materials and labour instead of sourcing them thousands of miles away from their base. They care about the preservation of local traditional skills. Fourthly, companies' aesthetic considerations combined with quality of manufacture tend to be more durable, thus guaranteeing that garments can be worn over much longer periods of time. Finally, financial viability is secured since companies can charge more and customers are willing to pay more because they know that these fashion products are made in a fair practice ecosystem.

Slow fashion can be related to craft as by nature craft work is not subject to time restrictions. The word handicraft and its derivative craft imply that a crafter produces an object by hand. Even machines in the production of handicrafts are in the service of the crafter [27]. Craft can be defined as the design and manufacture of an object by the same person who exercises full control over all processes [3]. Thus, the crafter imbues his/her manufactured objects with his/her personality, and also with the crafter's creative, expressive and authentic values. Pleasure in making derives from the fact that crafters take quality time to complete the task, and simultaneously immerse themselves in thoughts and feelings about how to improve their craftsmanship skills, which is also reflected in the quality of the object produced [27, 29]. Tsoumas [29] also observes that there are key differences between design and craft where in craft the manufacturing process from conception to execution is carried out by the crafter or a small group of crafters. Thus, in the craft practice, there is no division of labour between designers and manufacturers as in industrial production. Another important aspect of craft practice is lack of reproducibility as crafters tend to improve constantly in order to expand the techniques they apply to the artefacts they create. This makes each artefact unique as there are no exact copies because spontaneity and improvisation dominate in the making process ([29, 27]).

2.3.1 Tacit Knowledge Transfer and Knowledge Sharing

Organizational knowledge creation theory focuses on the creativity of firms, change, and innovation. Related to organizational knowledge, the social practice view has prompted scholars to analyse existing, tightly-knit groups operating in socially stable corporate settings where individuals are able to access to tacit knowledge through socialization in practice. In this concept, tacit knowledge, since its introduction into organization science serves mainly two purposes namely, the foundation of social practice, and the foundation for innovation [20]. Individuals acquire tacit knowledge or learn to engage in practical activities by becoming participants in social practices under the guidance of people who are more experienced than they are. A social practice can be formed by any coherent, complex, coordinated form of human activity where goods practices internal to this form of activity are realized and extended when crafters strive to achieve standards of excellence (Tsoukas, 2003). Virtuous practices are an internal good for a crafter, musicians and so on. For example, becoming a virtuous master tailor is an internal good as it supports and enhances the social practice of tailoring. Practitioners by being members of a social practice learn the "rules" of performance, skills, values, beliefs, and norms that forms their behaviour and also shape their practice [20, 29]. Practitioners' tacit knowledge (master tailors) consists of a set of particular abilities, such as dexterity in handling fabric and hand-stitching as tailors are instinctively aware that this constitutes the main focus in the production of a really good suit. Therefore, virtuous practitioners intuitively acknowledge which rules to follow. Simultaneously, they also intuitively recognize when the rules should be reformed or perhaps discarded when conditions emerge. Moreover, crafters become effective members of a social practice as this is formed by a complex network of people, the artifacts they produce, and the activities they perform [20].

One aspect of craft is that it can be applied in the production of luxury artefacts. The connection of craft to this of luxury is discussed below.

2.4 Craft and the True Meaning of Luxury

Craft consumption in advanced economies is more related to the luxury industry by connecting crafters to potential customers through personalization services. Personalization implies that customers specify their desires to the crafter who designs and manufactures the object usually at a higher price [3]. This establishes the true meaning of luxury as this type of luxury lies outside the so-called premium brands [12]. Kapferer and Bastien [13] sets out what they call the true luxury strategy away from the premium fashion business model. They refine the luxury business model and identify its fundamental parameters including the following parameters.

- Outstanding quality deriving from superior craftsmanship.
- Control of all production stages and strict requirements from suppliers who need to conform to the superiority of materials and trim used in manufacturing processes.
- Personalized services and customers' immersion in the retail experience.
- Aversion of discounts and other forms of sales at the retail level.
- Selective retail distribution due to the over-exposure of the house in retail markets impairing the aura surrounding the house as well as the artefact.

Moreover, [12] argues that a true luxury brand should not outsource its production to lower its cost structure as the country of origin is fundamental to the exclusivity of the goods as it emphasizes the importance of the national culture and psyche. There is also an ethical reason for retaining production within national borders. True luxury objects are the finest goods a country can produce and as such they can be appreciated by an international customer base.

Luxury's output in terms of retail operations is carried out in sumptuous retail environments replete with fascinating architecture and interiors which entice customers while interfacing with the luxury brand. Experiences in retail are discussed below.

2.5 Experiential Retailing

Experiences in consumption refer to evocative retail environments as customers immerse themselves into the shopping experience [21]. Businesses implementation of experiences relates to a novel way of marketing goods and services where marketers transform them into experiential cultural products whose value is increased in customers' eyes [9, 26].

Experiences correspond to a number of pleasure dimensions when customers engage with the brand experience. These include the sensorial dimension where customers are stimulated through their physical senses. The emotional dimension as moods and feelings experienced by the customer can be linked to the brand and its products or services, the cognitive dimension when customers think and evaluate their problem solving issues prior to purchasing, the pragmatic dimension relating to users' experiences, regarding the product's design as part of the human-artefact interaction, and finally, the lifestyle and relational experiences where customers engage in fitting their value system to the social context they belong to. The sensorial component is the most important dimension, but it is important to acknowledge that complex experiences have the capacity to engage customers with more than one pleasure dimensions [9, 26].

At the retail level which forms the core of the experience, customers enter an experiential store and immerse themselves in multiple areas which in turn enhance their experience with the brand [1, 21]. This explains why experiential brands in the luxury sector pay so much attention and expend vast amounts of money in flagship stores. Flagship stores or concept stores (in our case) demonstrate a strong communicative power in terms of symbolic connotations and customer traffic. This communicative power strengthens the brand identity and also generates awareness. Moreover, place specificities within a city are very important because customers flock to areas containing stores well-placed to promote the intangible elements of consumption in specific areas within the city [1].

3 Research Methods and the Case Study

The interpretivist methodology has been adopted for this study which subscribes to the logic of interpretation and observation in understanding the societal context of the world. This approach is integral to the qualitative tradition of research since it subscribes to the idea that knowledge is actively constructed by humans. It is focused on the contexts of how humans live and work which helps researchers to understand the cultural and historical background of people in specific societal situations. Moreover, researchers' own environment enables them to comprehend how their interpretation flows and is also influenced by their own personal, historical and cultural experiences [6, 24]. The researchers' intent for this study is to interpret a pattern of urban artisanal luxury phenomena in Taiwan which is configured by the actors who are involved in the research topic.

Thus, qualitative research methods have been adopted due to the nature of the aesthetic qualities that are constitutive of creative industries as these qualities are difficult to be captured by quantified data [6, 18]. Observation and documentation of visual materials have been employed in order to gain an in-depth understanding of how a creative micro-business functions and also by exploring the relevant aesthetic parameters in the production or consumption of this micro-business. The authors have conducted fieldwork in Taipei city, the capital of Taiwan between June and August 2020. The fieldwork included observation of events of a fashion firm, followed by 6 semi-structured interviews with participants including one interview with the owner of the fashion firm. Interviewees responding to a questionnaire have been asked how they proceed with their craft practice, material sources, production of small batches and on commission, the operation of the network, experiential retailing, business survival, inclusive growth within the community, and business viability and competitiveness. The authors have also randomly held informal discussions with three customers who happened to attend a reception event at the store in Taipei as these informal discussions have been very instructive and also complemented the data gathered from the semi-structured interviews. Lastly, data triangulation has included trade publications in Taiwan as well as in France and also critiques on the collections by art and design critics in France following relevant exhibitions in public exhibition spaces in France. It should be noted that the fashion firm has also established operations in France through a shop in Paris complemented by a wholesale showroom which covers the country as well as the rest of Europe.

It should also be noted that the case study is an atypical example of the Taiwanese textile and clothing sectors as Taiwan in these sectors overwhelmingly forms part of global supply chains dominated by well-known global fashion brands. The major output of the sectors is directed towards functional materials where Taiwan is a world leader. Manufacturing operations in garment factories is performed mainly by the same textile mills which produce the fabrics and these are dispersed in the Asian, African and Latin American Continents. This as it is understood, is in contrast to handcrafted clothing pieces and small batch production performed locally, using luxurious materials and trim sourced in Europe as well as in China. Thus, this particular type of manufacturing and retailing forms only a small albeit significant fraction of the sectors. It has been a considerable endeavour for the authors to discover craft related fashion businesses in Taiwan. However, this admittedly small subsector represents a refreshing departure from global operations and the ubiquity of luxury fashion brands which dominate global markets.

3.1   The Case Study

The case study has involved an exclusive micro fashion firm, Sylvie Creations owned and run by a legendary Taiwanese designer who has had decades of experience in crafting couture pieces on commission in her workshop as well as ready-to-wear collections produced through collaborations with small manufacturing workshops scattered in the Taipei area. The designer, Sylvie has established a loyal following of customers in Taipei while in France she is a well-respected member of the Paris fashion scene. She has also been decorated in France with the National Order of Merit for her services to the fashion industry. At the time of the interviews, the designer was fully engaged with her autumn 2020 collection.

Sylvie has stated that she has been in the fashion business for a long time and she managed to survive because she was consistent in maintaining her vision of exclusivity and using the best materials available. This vision led her to France where, as an apprentice, she perfected her skills in pattern cutting and sewing techniques. Upon her return to Taiwan, she opened a design studio and workshop where initially she was designing occasion wear for individual customers as commission pieces. Some of her customers have developed a loyal following whom she considers them more as friends because they share the same appreciation for the couture craft and the materials used.

In terms of design and pertinent design processes for the ready-to-wear collections, the authors have interviewed Sylvie's chief designer who had begun her fashion career with Sylvie as an apprentice. Following her apprenticeship, she has been employed on full-time basis and hence developed a profound understanding of Sylvie's vision in terms of design, selection of fabrics and the slow fashion mind-set in manufacturing.

In the interview she stated that design processes are more like reiterations of shapes and volumes which have been in Sylvie's mind for years, but these do not represent nostalgia or repetition but rather timeless pieces which evolve gradually. Every season these shapes and volumes are reinterpreted into new collections using diverse materials which create freshness in all the collection pieces. Some silk fabrics are hand-painted by local artists who have been collaborating with the company for years, while other pieces are hand-embroider so every season customers are attracted by these subtle details which bring newness to the collections.

Sourcing materials and trims is carried out months before production as it takes time for highly specialized small fabric rolls to be woven, dyed, and finished to specific colours. In the interview, the sourcing manager explained that all fabrics and trims that the company uses come from sustainable sources. Especially for the higher end pieces of the collection as well as the exhibition pieces currently displayed in a temporary exhibition in a French museum, the company has been collaborating for years with a one of a kind textile mill in Southern China as they revived a technique of dyeing and finishing silk fabrics which goes back to the Tang Dynasty. This fabric production method involves silk yarns which are hand-woven by crafters. Weaving is followed by a lengthy dyeing process where dyed fabrics are covered in mud and let to dry. After drying, workers wash the mud off and let the fabrics dry again in grass fields until they achieve their vivid colours. In terms of the ready-to-wear collections, the company sources fabrics from France and trims from Taiwan like fastenings using silk yarns. Chinese fastenings involve a slow-making process as all fastenings are made by a number of ladies who have been collaborating with Sylvie for years and they braid the fastenings by hand.

Production is carried out in highly specialized garment workshops in Taipei as well as the designer's studio where one-off couture pieces are made. The production manager interviewed has explained that his job involves the coordination of production, negotiating prices and allocating production batches to a number of workshops around the Taipei area in a web like formation. Samples made in the corporate sample room are all commission pieces. However, the rest of the production is outsourced to workshops where the workshop owners have developed a cordial relationship with the company for years as quality levels are kept to the highest level. Production prices are well above average, as both parties acknowledge that high quality work will bring more customers and retain existing ones. Hence, quality levels are maintained.

In retailing, the manager of the store in Taipei which is the company's only retail outlet has pointed out that the customer base is principally mature ladies who can appreciate superior quality and craftsmanship. The goods tend to be on the expensive side, but customers who visit the store understand and appreciate superior materials, the complexity of manufacturing and the hand-made work involved in most of pieces in the collection. Special events are organized throughout the year especially at the beginning of each season where customers are invited to attend a concert of chamber music where there is also a reception with refreshments. Customers can browse the collections, discuss the design details with the retail staff who are knowledgeable and helpful, thus facilitating customers' purchasing decisions. In terms of retail atmospherics, the shop has been decorated in a neo-bohemian style which mixes of European as well as Chinese elements such as antique furniture pieces from France and also some antique fittings from Taiwan. These fittings add character to the shop and also help merchandisers to display the collections in an appealing manner.

During the interview, Sylvie has stated how her slow fashion mentality enhanced the quality of life and how her micro-business remained competitive and contributed to inclusive growth. She explained that the brand has no reason to demand tight time restrictions from the artists who paint some of dresses and coats or the ladies who make the silk fastenings or the weavers and dyers. By nature these processes are slow, thus demanding a certain natural rhythm. This is acknowledged also by the customers as they understand that it takes time to make a beautifully hand-crafted garment. This commitment to slowness can also be reflected in an improved quality of life not only for the customers but also for the company's employees. Sylvie has been running a successful business for more than thirty years and she has never thought of expanding the company beyond her capabilities by producing larger collection volumes at cheaper price points in order to grow. Her principle has always been adherence to small quantities making use of superior materials and manufacturing methods which enhance the durability of all clothing items in the collections as they are often passed on to next generations as family heirlooms.

Finally, following the autumn reception which the authors attended in the shop, they held a number of informal discussions with some of the customers, one of them said:

"…I know Sylvie for a long time…I came here first time to make an outfit for a dinner party and since then I am a regular customer…I attend most of her presentations featuring her new collections every season and as you can see the atmosphere here is delightful…I have also brought friends who became customers and I have also made a few friends here as I believe that we all share the same appreciation for craftsmanship and exclusive materials… yes they are expensive but you know something, I'm still wearing most of them regularly as they don't date, they are not fashionable and loud they are stylish…" (Anonymous customer, Informal interview, 23rd August, 2020).

4 Discussion

Aesthetic capitalism can be included in the equation of ethical local production and responsible consumption as this has been revealed in the case study. More specifically, Sylvie Creations has demonstrated a certain ethos which has been set by the owner/designer from the outset. This business ethos has permeated all decisions with regards to both the strategic orientation of the firm as well as its operational processes from design, sourcing and manufacture to retail and also extended to customers who are loyal to and value the firm's creations over the long term. This ethos conforms to aesthetic capitalism's principles regarding the supply of fascination not in terms of a "wow factor" mostly associated with global luxury brands but in a more understated manner which we refer to as "cultured" as the fashion firm does not make use of glamorous campaigns or sells in glitzy retail environments. On the contrary, Sylvie Creations has intentionally chosen to remain under the radar of the fashionistas scene as it focuses more on the quality of its production coming from the materials used and the craft involved in manufacture, in combination with the subdued but no less evocative retail atmospherics of its store. Moreover, the notion of a metropolitan lifestyle prevails in Sylvie Creations as the firm is part of the capital's (Taipei) culture calendar for the initiated and highly sophisticated customers who frequent the shop and follow its activities on a seasonal basis. The culture of markets can also be confirmed as Sylvie Creations excels in social relations, and this has built a structure of meanings around the fashion business. These meanings and social relations are interpreted as meaningful by the customers who acknowledge the stylistic endeavours and aesthetic qualities of the firm's collections. These aesthetic qualities are complimented by the retail environment where the sensorial dimension is evident from the antique fixtures to retail atmospherics where customers immerse themselves in the experience as this is ultimately translated into sales.

In terms of the viability and competitiveness of the firm, network resources are present in a relational manner which includes the artists and crafters who paint, embroider, make trim for the collections and make the garments, thus, generating uniqueness which is not easily replicated by competing firms. Therefore, it can be argued that craft can be a valuable resource which can enhance competitiveness. Other network resources are also present as these are reflected in the adherence of Sylvie Creations to slow fashion principles. These include the use of local crafters; good salaries for all its workers and partners; the preservation of traditional skills, the superior quality and durability of its garments in any given collection; and the timelessness of its design which is not fashionable but stylish. Fabric sourcing from France is perhaps the only element which contradicts the slow fashion tenets, but in this case it is considered an unavoidable necessity as France still produces some of the world's best couture fabrics. Another crucial parameter is that Sylvie Creations imbues all collections with a definitive sense of local aesthetics which confers a differentiation advantage for the firm [23]. Moreover, the firm conforms to the fundamentals of the true meaning of luxury (which also reinforces its competitiveness) where outstanding quality has been achieved by full control of all manufacturing processes from materials procurement to manufacture and trim under the watchful eye of the sourcing and production managers. Personalization also forms a strong part of the business as in occasion wear where loyal customers return regularly. Personalization also can be another point of differentiation advantage which according to [23] differentiation can be based on the superiority of some fashion firms to manipulate and manage their image and their product ranges.

Ethical considerations such as those which have been presented in the case study can contribute to sustainable competitiveness and assist in constructing a healthy urban economy. The fashion firm's characteristics can be superimposed in Fig. 1 shown above. More specifically, Sylvie Creations' network configuration concurs with most of the parameters inherent in the diagram. These include the creative class which refers to all participants in the network from the crafters to the artists including of course Sylvie herself; the skills and labour quality are reflected in the superior methods employed in the manufacture of the collections; the network's sociality is also present as it can be related to aesthetic capitalism's culture of markets and more particularly in its social structure dimension. The culture of markets is also evident in the production of meaningful collections imbued with meanings while the sophistication of demand refers to Sylvie Creations' customer base who are part of Taipei's metropolitan lifestyle thus, their sophistication is reflected in their fashion-related purchases. The only element which has not been adequately explored empirically is the infrastructure of cultural institutions. Nevertheless, Sylvie herself, as well as her chief designer graduated from a renowned Fashion College in Taipei which has been at the forefront of the fashion scene for decades.

5 Concluding Comments and Perspectives

The urban artisanal luxury network in its cultured configuration does not form part of the ostentatious strategies pursued by global luxury brands. It becomes a significant setting where a local economy can thrive following all ethical considerations from the slow fashion movement and the tenets of true luxury to urban inclusive growth. In the case study, the authors have found no evidence that aesthetic capitalism cannot be symbiotic with ethics, sustainable competitiveness and inclusive growth within an urban conurbation. Combining all the concepts outlined above, the authors devise a framework for a cultured artisanal luxury network in fashion (see Fig. 2).

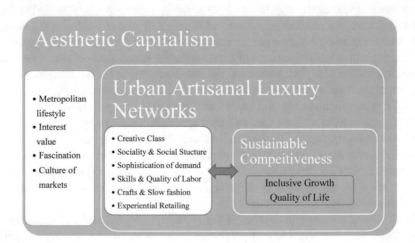

Fig. 2 The relationship between aesthetic capitalism and artisanal production leading to sustainable competitiveness in an urban network. *Source* The Authors

Research limitations include the fact that the research has been based on a single case study. More case studies from Taiwan's other cities could be explored, especially in the old city of Tainan where the authors are aware of a number of firms which produce hand-made artefacts such as bags directed to local and also tourist markets. Moreover, the authors acknowledge that artisanal production of artefacts can be combined with sustainability in upcycling materials by transforming them into new pieces of clothing and accessories. This topic has recently generated a great deal of interest in the fashion industry. Upcycling may include higher-end garments and can be an expensive exercise as it is a laborious and time consuming manufacturing process, demanding skillful cutting and stitching dexterity.

Lastly, more research on the matter can also be extended to other countries in the Asia–Pacific region where it is assumed that artisanal production takes place and more cases could be explored in order to substantiate the argument of craft practices combined with competitiveness in urban conurbations thus, reinforcing the UN's SDGs of improving the quality of life in the city.

References

1. Arrigo E (2011) Fashion, luxury and design: store brand management and global cities identity. Symphony Emerg Issues Manag 1:55–67
2. Billbao Osorio B, Dutta S, Lanvin B (2013) The global information technology report 2014. Geneva: World Economic Forum
3. Campbell C (2005) The Craft Consumer. J Consum Cult 5(1):23–42
4. Coe NM, Dicken P, Hess M (2008) Global Production Networks: Realizing the Potential. J Econ Geogr 8:271–295
5. Cooke P, Lazzeretti L (2008) Creative cities: an introduction Cheltenham: Edward Elgar Publishing Limited
6. Creswell JW (2009) Research Design: Qualitative, Quantitative and Mixed Methods Approaches, 3rd edn. Sage Publications Inc., California
7. Dyer JH, Singh H (1998) The relational view: cooperative strategy and sources of interorganizaitonal competitive advantage. Acad Manag Rev 23(4):660–679
8. Fletcher K (2008) Sustainable fashion & textiles: design journeys. Earthscan, London
9. Gentile C, Spiller N, Noci G (2007) How to sustain the customer experience: an overview of experience components that cocreate value with the customer. Eur Manag J 25(5):395–410
10. Godart F (2018). Culture, structure, and the market interface: exploring the networks of stylistic elements and houses in fashion poetics. https://doi.org/10.1016/j.poetic.2018.04.004
11. IPBES (2019) Global assessment report on biodiversity and ecosystem services of the Intergovernmental Science-Policy Platform on Biodiversity and Ecosystem Services. Bonn, Germany
12. Kapferer JN (2012) Why luxury should not delocalize: a critique of a growing tendency. The European Business Review, March-April, 58–62
13. Kapferer JN, Bastien V (2012) The luxury strategy, 2nd edn. Kogan Page, London
14. Kitson M, Martin R, Tyler P (2004) Regional competitiveness: an elusive yet key concept? Reg Stud 38(9):991–999
15. Knoben J (2011) The geographic distance of relocation search: an extended resource-based perspective. Econ Geogr 87(4):371–392
16. Lavie D (2006) The competitive advantage of interconnected firms: A extension of the resource-based view. Acad Manag Rev 31(3):638–658

17. Lorentzen A, Jeannerat H (2013) Urban and regional studies in the experience economy: What kind of turn? Eur Urban RegNal Stud 20(4):363–369
18. Miles MB, Huberman AM (1994) Qualitative data analysis, 2nd edn. Sage Publications Inc, California
19. Murphy P, De la Fuente E (2014) Aesthetic capitalism. Bill Academic Publisher, Boston, Netherlands
20. Nonaka I, von Krogh G (2009) Tacit knowledge and knowledge conversion: Controversy and advancement in organizational knowledge creation theory. Organ Sci 20(3):635–652
21. Pine J, Gilmore J (2013) The experience economy: past, present and future. Sundbo J, Sorrensen F (eds), handbook on the experience economy. Cheltenham: Edward Elgar Publishing Ltd
22. Porter ME (1998) Competitive advantage: creating and sustaining superior performance. Free Press, New York
23. Richardson J (1996) Vertical integration and rapid response in fashion apparel. Organ Sci 7(4):400–412
24. Ritchie J, Lewis J, McNaughton Nicholls C, Ormston R (2013) Qualitative research practice: a guide for social science students and researchers. Sage Publications, Los Angeles
25. Rosenthal SS, Strange WC (2004) Evidence on the nature and sources of agglomeration economies. In Henderson H, Thisse JF (eds), handbook of regional and urban economics, Vol. 4. London: Elsevier
26. Schmitt B (2011) Experience marketing: concepts, frameworks and consumer Insights. Marketing 5(2):55–112
27. Sennett R (2008) The craftsman. Yale University, London
28. Thomas D (2019) Fashionopolis: the price of fashion and the future of clothes. Penguin Random House, London
29. Tsoumas J (2013) The idea of handicrafts and the modern design formation: Coincidences and failures. METU J Fac Archit 30(2):55–62
30. UNESCO (2016) UNESCO creative cities. Senate of Economics, Technology and Research, Berlin
31. United Nations (2015) Transforming our world: the 2030 agenda for sustainable development (resolution adopted by the general assembly ed., pp. 1–35): United Nations

Printed in the United States
by Baker & Taylor Publisher Services